The
Underground Guide
to
Las Vegas

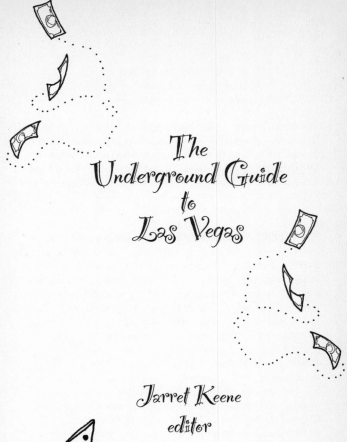

The Underground Guide to Las Vegas

Jarret Keene

editor

Manic D Press
San Francisco

Dedicated to Matt O'Brien
the Last Alt-weekly Editor on Earth
and a true underground explorer

Thanks to Princess Anne, Seth Flynn Barkan, Gregory Crosby, Scott Dickensheets, Dr. Joshua Ellis, Izzy Fizler, and Jennifer Prosser for making this book happen. Special thanks goes to Dayvid Figler, Russ Cannon, Kevin Capp, Bill Hughes, Matt Kelemen, Robert Kimberly, Al Mancini, Kelle Schillaci, Mike Zigler, and the whole crew at *Las Vegas CityLife*.

And thanks to my wife Jennifer: I'm lucky in love.

EVER-LOVIN' DISCLAIMER: Just as almost everything in life is negotiable, so too everything is conditional and SUBJECT TO CHANGE without a moment's notice. A listing in this book does not imply endorsement. All opinions are those of the individual authors, and not necessarily those of the publisher. All information is allegedly accurate as this goes to print but, hey, deal with it, okay? If you find something here that just ain't so, please be kind enough to let us know.

Cover: Scott Idleman / BLINK
Illustrations: Scott Dickensheets

ISBN 0-916397-99-8
Printed in the United States of America

- Contents -

ABANDON ALL HOPE,
YE WHO ENTER!¹

God and Satan loathe Las Vegas. Why? It makes them both obsolete. Here in Sin City, you'll encounter heaven and hell, angels and sinners, lovers and liars. Sure, sometimes it gets difficult to tell the difference. But that's half the fun. The other half is figuring out just where these creatures are taking you. Maybe it's the airy heights of previously unimagined pleasure. Maybe it's just another gutterful of stars sluicing its way toward a sulfur-stinking sewage pit.

Vegas, baby.

Whatever the outcome, this town is all about removing you from the everyday world and plunging you into pools of translucent desire, greed and lust. Yet at the same time, Vegas will dazzle you with its hospitality, friendliness and, ultimately, its honest mission to make you *feel* like you're being entertained.

True, those towering hotel-casinos encased in weird neon weren't built on the backs of winners. Rather, they were constructed on the piles of huge financial losses that gamblers never fail to suffer by the time they get on a plane back to Pocatello or wherever the hell it is they come from.

In spite of this, Vegas must be doing something right, since people arrive here—and almost always return here—in droves. The local economy bustles along like the last of America's boomtowns. Indeed, what Vegas does right is easy to articulate: it creates a powerful and unique illusion that you're having fun. And for many people, illusions can be just as crucial as reality. In fact, many marriages, relationships, deals, treaties and other arrangements rely on an illusion to keep them from falling apart. Just ask any of the characters in Eugene O'Neill's *The Iceman Cometh*: without pipe dreams (like the dream of winning that giant hand in a poker game), people's souls wither and die.

The Underground Guide to Las Vegas isn't here to kill off your soul, of course. Or destroy your flesh. Instead, this book seeks to help you see behind the neon veil and into something a little more substantive in the Vegas scene. If you plan on coming to Vegas with more than the mainstream experience in mind, do yourself a favor and read this book cover to cover.

There are many of you out there looking for something a bit different, something outside of the run-of-the-mill, rock-'em sock-'em, hotel/casino experience on the Strip. Instead, you're a member of that rare breed of traveler who wants to "do Vegas," but not in a cheesy way. You wanna get your hair mussed, your shoes scuffed, check out some art, music, thrift stores. You wanna return home completely penniless and proud of yourself, wearing a smile that says, "I survived the amoral vortex more commonly known as Vegas … and all I got was the shirt stolen off my back." Okay, well, maybe not all that.

Seriously, if you're reading this, you're committed to seeing a side of Vegas that nobody—not the casino corporations, not Las Vegas Metro—wants you to see. Why? This book takes you into the sparkling depths of the underground, where the coolest sights and sounds, hippest bars and eateries, worthiest galleries and madams await. So you won't have to go broke or lose your mind taking friends and family to see the same old impersonators, goofy magicians and threadbare comics. (Hell, you can just head to Atlantic City if you enjoy that kinda stuff.) Instead, here's everything you really wanted to know about Vegas, but were too afraid to ask...

Jarret Keene

Need a **free map** of Las Vegas? Visit the Las Vegas Convention Authority online: www.lasvegas24hours. com/press/vegas_maps.html. Also, *Monorail* magazine offers a free map of the Strip (but not downtown Vegas). Find it at Monorail stops and Strip hotel-casinos.

-----

1. Inscription at the entrance to Hell, from *The Divine Comedy: Inferno* by Dante Alighieri

EMERGENCY INFORMATION

Here's the righteous info everyone should have handy at all times, for whatever reason.

Dial 911 from any telephone for police, fire department or an ambulance.

For non-emergency assistance from Las Vegas Metropolitan Police Department, dial 311 from any telephone. (An "emergency" is any situation in which property or human life is in jeopardy and prompt summoning of aid is essential.)

Suicide Prevention Hotline: 1-877-885-HOPE

Gambling Addiction Hotline: 1-800-522-4700

Alcoholics Anonymous: 702-598-1888

Rape Crisis Hotline: 702-366-1640

AIDS Hotline: 702-474-2437

Clark County Poison Control: 1-800-222-1222

Alcohol-Drug Treatment Referral: 1-800-454-8966

WestCare Detoxification Center: 702-383-4044

WestCare Youth Emergency Shelter: 702-385-3332

Clark County Child Abuse Hotline: 702-399-0081

Senior Protective Service: 702-455-4291

Domestic Violence Hotline: 702-564-3227 or 702-646-4981

Planned Parenthood: 702-878-7776

24-Hour Veterinarian: 702-876-2111

AAA Emergency Road Service: 1-800-400-4222

University Medical Center, 1800 W. Charleston Blvd., 702-383-2000. Hospital Emergency Room: no medical insurance necessary.

UMC Quick Care: 1769 E. Russell Rd., just east of the Tropicana Hotel, 702-383-3600. Walk-in clinic.

Free/Low-Cost Medical and Dental Clinics:

Clark County Health Dept. Adult Clinics: 625 Shadow Lane, 702-383-1303, M-F 8:30 a.m.-4:30 p.m. HIV, STD, and Family Planning clinics. Walk-ins okay but appointments recommended.

Free Healthcare Project: 600 Whitney Ranch Dr., Suite D20, Henderson, 702-214-4713. Free healthcare screenings every Sat., 9 a.m.-2 p.m. Call for more info.

UNLV Dental Clinics: The University operates four low cost dental clinics in the city. Call 702-647-1074 for information on how to get your teeth cleaned and fixed cheaply.

COMING & GOING

& GIMME SHELTER, TOO

Southwest Airlines remains the cheapest way to get in and out of Las Vegas from most major cities. Yes, Southwest is a little gnarly, what with the whole first-come/first-served seating policy, no meals and cramped quarters. But if you're looking for the most efficient and inexpensive method, you can't beat Southwest. Good luck finding tickets on websites like **Orbitz.com**, **Travelocity.com**, **Expedia.com** or **LowestFare.com**, because they don't include Southwest flights. (Just in case things change, you should at least check 'em out.) Go to **www.south west.com** or check with your travel agent. To make sure you're getting the best possible deal, check Southwest's prices against **United** (www.united.com) and **America West** (www.americawest.com)—sometimes these airlines sneak in some cheap tix.

It's true that **JetBlue** (www.jetblue.com) offers $99 one-way tickets to Vegas from New York. But unless you live in or near the Big Apple, this won't help. Another discount carrier worth checking out is **Song** (www.flysong.com).

For last-minute, gotta-get-the-hell-outta-Dodge tix, Southwest won't do you much good, though, so we don't recommend taking a last-minute flight to Vegas. This is a town that needs to be approached carefully, thoughtfully, with a certain amount of precaution. That way it's a lot more fun when you destroy your brain with alcohol, gambling and sex.

Other options are more scenic, of course. Sadly, **Amtrak** (1-800-USA-RAIL, www.amtrak.com) doesn't reach as far as Vegas, so ultimately you'll have to **Greyhound** (800-229-9424; www.greyhound.com) it from L.A. Going Greyhound is cool if you're planning to write a novel about degenerate souls who seek further degeneration in Sin City. Otherwise, we don't recommend it for the basic reason that you don't save any money these days taking Amtrak or Greyhound, and the sleeping quarters will, if anything, be a tad tawdrier than Southwest's cramped, brown-clothed bucket seats.

And it doesn't matter when you get here: winter, spring, summer or fall—Vegas is consistently packed, and there's always a holiday or huge convention in town to keep the Strip bustling every weekend. So plan a trip around your schedule and not the other way around.

Besides, the Vegas underground is rarely affected by anything like the National Finals Rodeo.

Getting Around

Once you arrive at **McCarran International Airport** (5757 Wayne Newton Blvd., 702-261-5211), do not—repeat: do not—stop to pull any of the one-armed bandits that greet you as you step off the plane. The slot machines are put there for a purpose, and that purpose is to fleece your pockets before you even make it to the baggage claim. Ignore them, walk right past and get your bags.

It'd be a good idea, though, to buy a big bottle of water and some eyedrops before leaving home, especially if you're here during the summer months. See, Vegas is all about really, really dry heat that will cause your throat to hurt, your nose to bleed, and your eyes to itch. It's like being on the surface of Mars. Or rather like being inside an active blowdryer. Oh, and secure sunscreen, sunglasses and a map of Vegas upon arrival. You don't want to end up at the **UMC Quick Care** (1769 E. Russell Rd., just east of the Tropicana Hotel, 702-383-3600).

Taxis are criminally expensive here in Vegas, and a few bad-apple drivers have been caught "long-hauling" (or ripping off) tourists. So use taxi service sparingly. If you're in a group, a $15 cab ride won't hurt as much, but you still should get your bearings. The airport runs perpendicular to the south end of the Strip. If you're heading south on the Strip after having been driven north on the interstate, you're being long-hauled, and you need to report your driver to Metro.

The interstate isn't always a bad sign, though. If

your driver starts at the south end of the Strip to ferry you to your sin bunker, located on the far north end of the Strip (like, near the Stratosphere), chances are you're being long-hauled. Remember: Avoid the Strip if you're in a hurry or need a cheaper cab ride, 'cause the Strip is almost always bumper-to-bumper traffic at any time, day or night. So if you ain't gonna cruise it, lose it.

An airport shuttle to the Strip won't be much different pricewise: $20-$40 for one to four passengers. But individual private shuttle rides and limos have a flat flee of $5 to a Strip hotel, $6 to a downtown hotel and $7 to an off-Strip hotel—and they're about $35 for up to six folks. For a complete list of ground transportation services from the airport to the Strip, go to **www.mccarran.com**. Or try one of the following: **Bell Trans** (www.bell-trans.com, 702-739-7990), **Grayline/Coach USA/Express** (702-739-5900), and **ODS** (702-876-2222). Or try one of these limo companies: **C.L.S.** (702-740-4040), **Las Vegas Limousine** (702-736-1419), and **Showtime** (702-261-6101).

There are two bus routes to and from the airport, courtesy of **Citizen's Area Transit** (www.rtcsouthern nevada.com/cat). First, there's **Route 108**, which runs the length of the Strip (a.k.a. Las Vegas Blvd.) and reaches all the way to downtown. You can grab one every half-hour from 5 a.m. to 1 a.m. And then there's the 24-hour **Route 109**, which also reaches downtown via Maryland Parkway, a street that goes right through the university district. A dollar is the standard fare for a one-way ride off the Strip; it's a $1.50 on the Strip. Buses get really packed during the peak hours. The farther you are from the Strip, the less likely your routes will operate past midnight.

Renting a car in Vegas is pretty reasonable and makes sense. After all, Vegas is nothing like a real city, since you can get around easily enough by car (as long as you're not driving on the Strip at night!). And besides, all the car rental companies are located at the airport, making it easy to pick up and drop off a car. **Alamo** (702-263-8411 or 1-800-327-9633), **Allstate** (702-736-6147 or 1-800-634-6186), **Avis** (702-261-5595 or 1-800-831-2847), **Budget** (702-736-1212), **Dollar** (702-739-8408 or 1-800-800-4000), **Hertz** (702-262-7700 or 1-800-654-3131) and **National** (702-261-5391 or 1-800-227-7368).

There are trolleys in Vegas! **Las Vegas Strip Trolleys** (702-382-1404) are replica streetcars that travel up and down the Strip … very slowly. Every time I've hitched a ride on one, I've gotten nowhere fast. And the claim that a streetcar arrives every 15 minutes is nonsense. Still, if you're in no hurry and enjoy sweating with strangers in stop-and-go traffic for long periods of time, then by all means pony up the $1.30 fare. Children are tortured for free.

A valuable resource during your Vegas vacation will be the new **Robert N. Broadbent Las Vegas Monorail** (www.lvmonorail.com). Sadly, it's not free, but it reaches 50 mph and glides high above the Strip, covering four miles—from the MGM Grand to the Sahara—in less than 14 minutes. It travels back and forth from one end of the Strip to the other seven days a week from 8 a.m. until midnight. You can buy your tickets ahead of time by going to the monorail's website. Highly recommended if you're gonna spend any time on the Strip. It's faster than a bus and cheaper than a taxi. The monorail is open 7 a.m. to 2 a.m. every day, and there are four types of fares: $3 per ride, $20 for 10 rides, $10 for a day-long pass and $25 for a three-day pass.

Selecting Your Very Own Sin Bunker

Vegas is easy to figure out, since it's laid out on a grid. For the purposes of investigating the underground, there's basically just the Strip (Las Vegas Blvd.), which runs south and right smack into the downtown area. There's also, of course, the southeast corridor called Green Valley, which borders the city of Henderson. (The **Green Valley Ranch** hotel-casino is where Michael Jackson, the king of pedophi … er, pop, often stays.)

If you're interested in staying in the heart of beautiful (ahem!) downtown Las Vegas, we suggest you check out Fremont Street, which is lined with many classic hotel-casinos like the **Golden Nugget** (129 E. Fremont St. and Casino Center, 702-385-7111, www.goldennugget.com), which has provided a setting for the reality TV show, "American Casino." We recommend this place without hesitation.

But if retro is your bag, there are some super-cheap retro-hotels in the downtown area, including the **Western** (899 Fremont St. at 9th St., 702-384-4620, www.westernhotelcasino.com), which offers $23.98 rooms. Built in 1965 and yet to be renovated, this is a real dump with cinderblock walls, garish lighting, cigarette-burned carpet/curtains/bedsheets. Yes, this is where junkies and the mentally deranged come to gamble away their few remaining pennies in the penny slots. In sum, the Western represents the underbelly of the Las Vegas dream. Small bonus: right across the street is a scary bar, **Atomic Liquors**. Avoid the food; it's truly horrendous.

Gold Spike (400 E. Ogden St. at 4th St., 702-384-8444, www.goldspikehotelcasino.com) will drive a wedge between your sanity and the world, located just a block from the Fremont Street Experience. The rooms are a lot nicer, but the casino is arguably a little smellier. Fewer lunatics on parade and more working-class folks. The lowest price here is $27, but go ahead and spring for a suite at $41. It's worth it. And forget about the barely identifiable railroad theme. Just be glad you're not staying at the Western. Penny slots! Don't eat here.

Built in 1941 (ancient history for Vegas), the **El Cortez** (600 Fremont St. and a block east of Las Vegas Blvd., 702-385-5200, www.elcortezhotelcasino.com), is the oldest standing hotel-casino in Vegas, and it's even closer to the Fremont Street Experience than the Western and the Spike. Still some down-and-outers on the casino floor but nowhere near as many as the previous bunkers. The food here is actually edible. At **Careful Kitty's Café**, there's a $1.95 breakfast (bacon and eggs with hash browns, toast and jelly), and certain sandwiches (The Player's Choice: meatball, Italian sausage, chicken or fish on a French roll) are just $3.95 and come with chips and cole slaw. Look for the hidden ice cream stand—it's tasty and cheap!

Built in 1955, **Golden Gate** (1 Fremont St. at Main St., 702-385-1906, www.goldengatecasino.net) has won a few awards (like "Best Downtown Hotel") from the local press. It's also "home of the Original 99-cent Shrimp Cocktail" (smothered in cocktail sauce). Standard room rates range from $35-$55 dollars per night. Whatever you do, don't shoot the real live piano players who perform on the casino floor from noon to midnight each day. They beat muzak hands down.

And **Lady Luck** (206 N. 3rd St. at Ogden, 702-477-3000, www.ladylucklv.com) isn't as charming as the others, given its more recent construction, but it's still pretty cheap.

Some pretty okay downtown hotels that give you plenty of bang for your buck include the **Plaza** (1 Main St. at Fremont, 702-386-2110, www.plazahotel casino.com), **Fitzgerald's** (301 E. Fremont St. with access at 3rd and Carson, 1-800-274-LUCK, www.fitz geraldslasvegas.com), and the **Las Vegas Club** (18 E. Fremont St. at Main St., 702-385-1664, www.vegas clubcasino.net).

Again, the **Golden Nugget** is the best of any of these—if you can find a room! (And remember that the price of all rooms everywhere goes up on the weekends and during special events.) It was remodeled in the '80s, but the use of gold trim and white canopies on its marble façade instead of harsh neon makes it stand out among its competitors. The rooms are the cleanest and most comfortable you'll find downtown, and they're nicer, in fact, than those in many of the Strip hotel-casinos. It'll cost you around $70 per night weekdays, $100 on the weekends.

If you insist on getting a hotel on the Strip proper, we hesitantly recommend the **Sahara** (2535 Las Vegas Blvd. S. at Sahara, 702-634-6411, www.saharahoteland casino.com) for a few reasons. First, it's cheap for the Strip: $40 per weekday and $100 per weekend night. Second, it's conveniently located midway between downtown and the south end of the Strip. Third, the new public monorail system stops there, so it makes for a great home base from which to conquer Vegas. Don't fall for the $7 buffet, though. Or anything else in the place, really. With the Sahara, it's all about location, location, location.

Also on the Strip is the **New Frontier** (3120 Las Vegas Blvd. S. between Spring Mountain and Desert Inn, 1-800-421-7806), which, in addition to housing the best redneck dive bar in the valley (**Gilley's**), offers $60 rooms Sunday through Thursday. (Weekend rates are always higher in Vegas.)

But if the Sahara doesn't sit right with you for whatever reason, try the clean $15 rooms at the **Laughing Jackalope Motel Bar & Grill** (3969 Las Vegas Blvd. S. at Mandalay Bay Dr., 702-739-1915). Located at the north end of the Strip and across from Mandalay Bay, the Jackalope is also a hop, skip and a jump from the airport.

Or you can stay at the **Artisan Hotel & Spa** (1501 W. Sahara Ave. at the I-15, 702-214-4000, www.the artisanhotel.com), where the cheapest rooms are $69.99. Location and convenience are what's great about this unique place, since it's just a few blocks from the Strip. Also, the place boasts reproductions of fine art in the lobby, halls and ceiling. Indeed, you'll be just seconds from the Strip and in the veritable lap of Lap-dance Row on Highland, home of some of Vegas's best grind shacks. Sadly, the food and drink here are prohibitively expensive, so plan on getting wasted and refueled elsewhere. If you have money to burn, everything on the menu is pretty damn tasty.

Or you can get your *Fear & Loathing* on at the evil-clown-plagued **Circus Circus** (2880 Las Vegas Blvd. S. between Sahara and Spring Mountain Dr., 702-734-0410, www.circuscircus-lasvegas.com), which in the last 30 years has transformed into a giant playpen for loud, demented tykes.

Yes, there's a neon-festooned **Motel 6** (195 E. Tropicana Ave., 1-800-466-8356) near the airport and the Strip for the budget-conscious, but the rooms are so sparely furnished and the services so minimal and the bar so non-existent that you might as well plan a vacation in Wichita.

There's also **Sin City Hostel** (1208 Las Vegas Blvd. S., south of Charleston Blvd., 702-868-0222, www.sincityhostel.com), which is located right on the Strip. And then there's a **USA Hostel** (1322 Fremont St. at 13th St., 702-385-1150, www.usahostels.com) downtown that has a swimming pool and a jacuzzi. They're both said to be clean, air-conditioned and Internet accessible—and they offer cable TV! Not bad for $16 a night. Don't really know much about the folks at USA Hostel, but the young folks who run Sin City Hostel seem friendly and helpful.

Staying off-Strip and away from the downtown area is a silly idea, simply because it's too far from the action—and from the monorails. But if you insist on keeping a little distance between the Strip and your sleeping quarters, try the French Quarter-themed **Orleans** (4500 W. Tropicana Ave. at Arville Rd, 702-365-7111 or1-800-ORLEANS, www.orleanscasino.com).

Staying for Good

You've been warned. It's all just ugly, rat-brain-on-meth, suburban sprawl here in Vegas. But if you insist, there's really only one area in town worth taking a look at: the **John S. Park Historic District**, which was added to the National Register of Historic Places in 2003. Named after a local banker and civic leader, the district is comprised of about 160 homes, some built as far back as 1931. More and more young professionals move here every month, but there's still a hardcore contingent of artists, musicians, and writers who find this area of town to be the most aesthetically pleasing and communal place to work. It's also located near the **Arts Factory**, a small yet vibrant hotbed of art galleries and renegade theater. There are still more than a few homes for sale or rent, so check it out now before prices become Californicated.

The best reason to live downtown is **First Friday** (www.firstfriday-lasvegas.org), a large arts and music gathering that takes place the first Friday of each month. See you there... after you run the bar gauntlet, of course.

UNLV's student union is a great place to look for cheap housing in the "real" Las Vegas. Fliers are everywhere — just grab one!

The UNLV website offers a free **Off-Campus Housing Locator Service** (http://ochl.nevada.edu/ochl) that anyone — even non-students — can use. And it allows you to organize your search to find specific information in these housing ads: rent, lease options, distance from UNLV, furnishings, occupancy limits, and student discounts. The service also allows you to sift through "roommate wanted" ads. To place a free ad you must be an enrolled student at UNLV.

If you're okay with living in a suburb far, far away from where the action is, check out www.lasvegas.craigslist.com. Or you can just call up any or all of the bohemian establishments (see "What Culture?" p. 180) and ask if anyone's looking for a roommate. Someone always is.

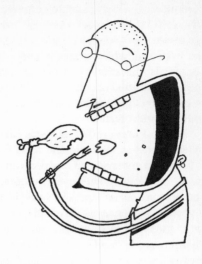

KICK-ASS CUISINE

AKA CHEAP EATS

There isn't a better place to eat well and cheaply than Las Vegas. After all, this town was made for one reason and one reason only: hedonism. That's right, we're here to help put the "ton" back into "gluttony."

Food in Las Vegas needs to be cheap and plentiful, really, because chances are, if you're reading this book, your budget is pretty tight. Here are some tasty and insanely cheap joints to get you fueled.

Downtown

Even though it's supposedly lower in calories, fat and cholesterol than beef, pork or chicken, goat meat isn't for everyone. But if you're a health-conscious and adventurous diner, you should definitely try **Birrieria Jalisco** (953 E. Sahara Ave. Unit E8 just west of Maryland inside Commercial Center, 702-892-9711). The meat is stewed, which makes it tender, and the terrific seasoning is what allows you to get past the gamey taste. Wrap it in a tortilla and season it with condiments and it's excellent, particularly when accompanied by a couple of cold Tecates.

Boston Pizza (1507 Las Vegas Blvd. S. just north of Oakey next to Olympic Gardens) is great because it's thick and greasy—and oh-so-cheap. A large, fully-loaded, all-the-way-with-everything-on-it pie (called "The Boston Special") runs $18, and it'll take two days for you and someone else to eat the whole damn thing. One of the city's best old-time pizza joints, this place also offers pretty good pastas, salads and wings. There are even a few Mexican dishes that'll work in a pinch. If you order "extra cheese," you'd better mean it.

Café Heidelberg German Deli (601 E. Sahara Ave. at 6th St., 702-731-5310) will make you beat on the sauerbraten with a baseball bat-sized fork, it's that good. I mean, does it get any better than a schnitzel sandwich served on a Kaiser bun with mayonnaise, lettuce and tomato, with German-style potato salad for eight bucks and change? Indeed, this Bavarian-style restaurant is run by German-born folks who know how to whet and then whip one's appetite. Or how about currywurst, a knackwurst served on a Kaiser bun with onions, curry ketchup and fries? Dinner is significantly more expensive, but if you've got 20 extra bucks in your pocket, try the Jaeger schnitzel, a pork cutlet battered and topped with sautéed mushrooms, onions and wine sauce. The goulash is so good you'll be dreaming of a Motherland you never knew.

Chicago Joe's (820 S. Fourth St. just north of Charleston Blvd., 702-382-5637) has been in business since 1975 in what was once clearly a private residence. The atmosphere here is somewhere between an actual Italian family's house and a scene out of "The Sopranos." The menu is pure homestyle Italian with pasta dishes running between $8 to $12, but you can add a bit of the exotic, like snails or mussels. Shrimp ($15) and lobster ($25) dishes are in the seafood section, if you've got money to burn, and the veal ($17) can be prepared parmigiana, picante, angelo or with mushroom and onions. Chicken ($14) comes in nearly the same varieties, and the sauces are all to die for. All meals come with a choice of house salad or pasta fagioli. The salad comes with a creamy garlic dressing that's just OK; the soup, on the other hand, is extraordinary. Among the highlights? Try the "Shrimp Joe."

Don't let the barrio look of the tiny **El Sombrero** (807 S. Main St. at Gass Ave., 702-382-9234) fool you. Built in 1950, this is one of Vegas's oldest restaurants. Menudo, anyone? No, not the Mexican pop group that Ricky Martin used to sing in. This is as authentic a Mexican joint as you're ever going to find in Vegas. Its downtown location attracts the city's powerbrokers: politicians, journalists, lawyers and (yikes!) alt-weekly writers. Lunchtime is always crowded, so if you want a booth you'd better arrive before noon or after 1:30. Otherwise, you may have to wait awhile for a place to sit. Speaking of Menudo, there's a jukebox chock-full of Mexican and American pop.

Tucked into a center at 10th and Charleston, **Doña Maria's** (910 Las Vegas Blvd. S. at Charleston Blvd.,

702-382-6538) has awesome tamales: mild pork in a red sauce, spicy chicken in a green sauce or the sweet pineapple kind. There are all the standard combos and fajitas, plus sections covering seafood and chicken platters. Sandwiches, T-bone steak and fried shrimp are offered for gringo souls who remain unconverted, while a children's menu focuses on what's proven to work for kids. Everything is available a la carte, and health-conscious diners can choose from an array of salads. Dinners range from about $9 to $15, while a special lunch menu offers filling combos for between $6.50 and $9.50. Mexican breakfast dishes are offered on weekend mornings, when the restaurant opens at 9 a.m.

Esmeralda's Café (1000 E. Charleston Blvd. at 10th, 702-388-1404) is where you have a pretty good chance of catching me and my wife on any given day or night. This joint is located a mere block from my downtown home, and it's simply one of the best cafes I've ever had the pleasure of dining at—period! Don't fall for the Mexican items; order the Salvadoran specialties, like the pupusas ($1.99), which are glorified meat-and-cheese filled cornmeal doughnuts and nirvana-inducing, especially when you munch 'em in between sips of Salvadoran beer (Bahia, Suprema). The pollo and bistec encebollado are out of this world, too. Got a cold? Try the caldo de pollo (chicken soup); it'll cure whatever ails you.

Florida Café Restaurante Cubano (1401 Las Vegas Blvd. S. and Wyoming, inside the Howard Johnson, 702-385-3013, www.floridacafecuban.com) doesn't just look, feel and sound like southern Florida, it also tastes like it. Sure, it can't possibly match up with some of the finest Cuban emporiums in Miami, but it still makes for *una fiesta sabrosa*. The appetizers are what you should order here, because they're all exquisite: corn tamale ($3.75), sliced fried sausage ($.395), fried pork skins ($3.25), croquetas or deep-fried balls of ham and cheese ($3.95), stuffed potato ball ($3.99) or a light and refreshing avocado salad. You can't go wrong. And if you've got a little spare change jangling in your pocket, splurge for the café con leche ($2.25).

Huntridge Drugstore Restaurant (1122 E. Charleston Blvd. at Maryland, 702-384-3737) is a throwback to the days of old when folks used to sidle up to the counter of a drugstore and order up some American classics, like Chinese food and a cheeseburger. OK, well, maybe not the Chinese food, but if you desire a bonafide retro experience in Vegas, then you could do worse than stepping inside this downtown time warp. Five or six bucks later, and you're full of yummy (if greasy) diner food.

Lotus of Siam (935 E. Sahara Ave., inside Commercial Center, 702-735-3039) is proud of the fact that *Gourmet* hailed it as the best Thai restaurant in the country—so much so that the staff clipped the article and mounted it on the anteroom wall. And why shouldn't the place be proud? It is, after all, the best Northern Thai you're gonna find outside of, well, Northern Thailand. That's because chef Saipin Chutima knows how to get your taste buds singing with just the right ingredients. The spicy sausages are served with cabbage, ginger and red onions, all of which help to cool your mouth in the wake of the well-spiced meat. As with nearly everything on the menu, the overwhelming spiciness of the meat is complemented by several more delicate tastes. The trick is finding them, given the overall spiciness that characterizes Thai food. Try tom yum kung, a hot and sour soup with shrimp ($4.95). Or go with the jung char num plar, raw prawns served with a spicy fish sauce, limes, fresh garlic and a sweet chili paste ($9.95). The lunch buffet here rocks!

Luv-It Frozen Custard (505 E. Oakey Blvd. at the Strip right next to Olympic Gardens Gentlemen's Club, 702-384-6452) is the best place in town for a frozen treat. Similar in taste and consistency to premium French vanilla ice cream, Luv-It's custard contains 10 percent butter fat content and an egg base that makes for a divine concoction. The menu says it all, really: Banana Fudge Krumble (hot fudge, sliced bananas cookie krumble), Desert (hot fudge, marshmallow, pecans) and Swiss Brownie Nut (Swiss chocolate, brownie, walnuts)—all of these liberally sprinkled on. Warning: This is a straight-up custard stand, so plan on standing there as you scarf. Or if you're driving, you should be able to eat it in your car provided there's a parking space for you. You'll never eat at Baskin-Robbins again.

Silvita's Mexican Grill (1236 Western Ave. at Wall St., near King and Charleston, 702-294-6100) is a cheap and plentiful place to eat. The breakfast menu has everything from pancakes and omelets to huevos rancheros and breakfast burritos. The lunch menu features chile rellenos, tamales, five types of tacos, tostadas, quesadillas, flautas, served with rice, beans and salsa for five or six bucks. The burritos are huge, ranging from $3.25 to $4.50. And an eight-buck combination platter will last you all day if you take a doggy bag back to your sin bunker. The tortilla soup—tasty and packed with black beans and corn—will make you think you've died and gone to heaven. Hell, you could make a meal out of the beans and some chips, this place is so damn good.

The first thing that strikes you when you enter **Tinoco's Bistro** (at the Arts Factory, 103 E. Charleston Blvd. at Casino Center, 702-464-5008) is the décor. The place plays up its Arts Factory location in a serious way. Tables are shaped like artists' palettes, surrounded by oversized booths. Artwork by local artists hangs on the walls. How's the food? The appetizer section ranges from crab cakes ($8.95) to chicken satay ($7.95) to calamari ($8.95). Try the corn and crab chowder ($4.95), and progress to lobster ravioli ($14.50).

University District

Al Basha (3969 S. Maryland Pkwy. at Flamingo, 702-699-7155) offers the best and cheapest all-you-can eat Lebanese lunch buffet that you'll ever wrap your taste buds around. Seven bucks gets you roast chicken, rice pilaf, stuffed vegetables and kofte (minced lamb brochettes), plus all the hummus, tabbouleh, pita bread and salad you can put down. If you don't make it in time for lunch, then stick around for some kabobs. In addition to the buffet, Al Basha offers eight salads ($4.95 to $6.95), about a dozen and a half appetizers ($3.95 to $8.95), more than a dozen entrées ($8.95 to $14.95) and a handful of vegetarian options ($5.95 to $10.95). Starters range from basics such as hummus, babaganoush and some extremely delicious dolma, to the more exotic Armenian pizza—which is actually just a tasty snack of pita bread spread with a thin layer of ground beef and spices, and folded into quarters. The moussaka is heavenly, consisting of layered eggplant, ground beef and potatoes covered with molten kafeletin cheese. It's a delicious and filling Greek version of lasagna that in many ways surpasses its Italian cousin.

East Boy Japanese Café (4755 S. Maryland Pkwy. at Harmon just north of Tropicana, 702-798-1777) has free Wi-Fi! So while you're surfing the Net for hentai sites, you can also fill your belly for less than a five spot with some tasty rice bowls: tempura, salmon teriyaki, beef sukiyaki or chicken teriyaki. If you've got six to eight dollars on you, you can splurge on some neatly arranged bento boxes that include things like seaweed salad, spaghetti (!) and tofu. Sounds weird, I know, but it's all very tasty—and very Japanese. Grab some Pocky (biscuit sticks covered with strawberry cream) on your way out.

Yeah, the **Freakin' Frog Beer & Wine Café** (4700 S. Maryland Pkwy. Suite 8, just north of Tropicana across from UNLV, 702-597-3237) has hundreds of beers available for your total and complete inebriation. But the food here is almost as good as the suds. The cheeseburgers are mouthwatering, and the chicken wings ain't too bad, either. The pizza's a little ho-hum, but really, that's the only weak item on the *very* affordable lunch menu.

Harrie's Bagelmania Bagel Factory & Restaurant (855 E. Twain Ave., 702-369-3322) beats any other bagel in town for sheer size and price. You can eat a giant-sized bagel for 95 cents (minus cream cheese) at a table, or you can get it to go for a mere 55 cents. For an extra 35 cents, you can haul off with an "everything bagel" or a "twisty double poppy seed." Be careful, though, these bagels are so big they practically bite back.

Wanna take care of breakfast for the week? On Tuesdays, Bagelmania has a special: a dozen bagels of your choice for $3.25. Just make sure and bring a forklift. And Monday through Friday, for a mere $4.50, you get three eggs (any style), meat (bacon, sausage or ham) and breakfast hash. Check out www.harriesbagelmania.com.

At **Paymon's Mediterranean Café** (4147 S. Maryland Pkwy. at Flamingo, 702-731-6030), the food runs the gamut from causal pita sandwiches (around $7) to fairly exotic Middle Eastern entrées ($9 to $11). For an appetizer, try the delicious flambéed cheese appetizer; spread it onto a pita and dip it into hot sauce. Where the menu really shines is in its entrée section. The kabobs ($10.95 to $12.95) and spinach pie ($9.95) are great, and the deeper you dig, the better it gets. Also sample the fesenjan ($9.95), tender chunks of chicken smothered in a sauce made of crushed walnuts. And when you're done, smoke a hookah in the adjacent Hookah Lounge.

The **Stake Out** (4800 S. Maryland Pkwy. just north of Tropicana across from UNLV, 702-798-8383) is where a lot of the UNLV art, drama, film and writing students hang out—and for good reason. This place has some of the cheapest, strongest drinks you'll find outside of the casinos, and the food is cheap and tasty. The $4 cheesesteaks—loaded with cheese, onions and green peppers—are pure heart attack, but you'll wish you had ordered two of them by the time you're halfway through the first one. Don't get fooled by all the other college grills near the university: this is the one to hit.

Tamana (2295 E. Tropicana Ave., 702-798-9595) has an outstanding menu that features an entire section of Gujarati vegetable offerings (all $7.99) below the dozen more familiar "vegetarian delights" ($6.99 to $7.99)—the former apparently being native to the western state of Gujarat. There are also 14 dosa selections ($4.99 to $7.99). Try a chicken tandoori wrap ($8.99) from the "specials" section: it's sort of like a small tandoori pizza, garnished with vegetables and a sweet garlic sauce, a delicious twist on an old favorite. Or opt for a tasty lamb vindaloo ($8.99) over saffron rice ($4.99). For bread, try aloo paratha ($3.99); it's chewy and very, very good.

North Las Vegas

Big Mama's Soul Food Rib Shack (2230 W. Bonanza Ave. at Rancho, 702-597-1616) will put a little South in your mouth with a three-dollar old-school barbecue sandwich called "The Killer" (not to be confused with Vegas-born, retro-glam rockers the Killers). And it even comes with cole slaw. Everything on the menu is cheap: fried chicken, ribs and Cajun standards like gumbo, jambalaya and shrimp creole. Big Mama's is an ideal place to grab takeout for a group of friends before heading to the Strip for a night of debauchery. And this food will stick to your ribs. Check out the menu at www.bigmamasribs.com.

Westside

The Omelet House (2160 W. Charleston Blvd. at Rancho, 702-384-6868) is just one of a few terrific breakfast nooks on the Westside of town. Obviously, the specialty here is omelets, which come in two sizes: regular and baby. (If you don't want to gorge yourself, we recommend the baby.) Some of the best selections include the "Bugsy Seigel," comprised of roast beef and Italian red sauce with jack cheese and sour cream. And the "Popeye" is a great spinach, mushrooms and jack cheese number that'll get you geared up to kick Brutus' (and Bluto's) ass and take Olive Oyle by the hand. Other terrific omelets include the "Rio Grande Surfer" (chorizo, onion and cheddar cheese), the "Green Hornet" (avocado, tomato and cheddar cheese), and the "Spartacus" (asparagus, mushroom, and cheddar cheese sauce). The pumpkin bread is good and spicy, and the "Famous Fried Zucchini" is spectacular, too. Come back a second time for the stuffed French toast. Too bad the coffee here sucks.

Not sure if **The Egg and I** (4533 W. Sahara Ave. between Decatur and Arville, 702-364-9686) is named after the 1947 comedy starring Fred MacMurray. But you won't be laughing once you sample this joint's first-rate eggs, pancakes and waffles. The coffee is much better here than other diners. The music? Not so good. More of a family atmosphere here, too, so prepare to deal with families and wailing children. Very busy on the weekends, but they're open for breakfast day and night. Stop in when things are slow.

The **Original Pancake House** (4833 Charleston Blvd. at Decatur, 702-259-7755) is the only place to go in Vegas for pancakes, people. Yes, it's a national chain, but God, is it good and fattening! The Apple Pancake with oven-baked granny smiths and cinnamon glaze. The Dutch Baby with whipped butter, powdered sugar and lemon. And the omelets! Try the mushroom omelet with sherry sauce. You can't go wrong here, no matter what you order. Just be prepared to do some serious eating, 'cause portions are super-sized.

Black Bear Diner (6180 W. Tropicana Ave. at Jones, 702-368-1077) is a bear-themed chain of diners that offers all the, um, bare breakfast necessities: omelets, scrambles, pancakes, waffles, etc. Try the California scramble, which contains avocado, onions, spinach, tomato, bell peppers and jack cheese. Then there's the Black Bear standard, which is two eggs, plus homemade Italian sausage, ham, bacon or corned beef. Or order the Hungry Bear and get all the meat you can eat. Don't like your menu choices? Build your own omelet then! On Sundays, some dude dresses up in a bear suit and makes the children laugh (and sometimes cry from fright). Pick up a bear mug, T-shirt or stuffed animal on your way out. (So family-friendly, in fact, you may want to lose your breakfast on the way out.) Lunch and dinner ain't bad, either.

Hue Thai's Sandwiches (5115 Spring Mountain Rd. at Decatur, 702-943-8872) offers some of the tastiest French bread subs you'll ever put in your mouth—for a mere $2.33 each. Yep, if this place served beer, it would have to change its name to "Heaven," the sandwiches—and the bread in particular—are that good and that cheap. See, when France colonized Vietnam decades ago, the French brought their breads with them. The two cultures initially had little understanding of each other's food, but soon they were able to merge and create a distinct blend of French cuisine and low-fat Vietnamese ingredients. It's called "Eurasian," and the house special sandwich (pork, cucumber, shredded carrot, lettuce, tomato and a green pepper) is the best. Feel free to brave any of the following: cold cut, Vietnamese meat loaf, shredded pork, shredded chicken, Chinese meat loaf, egg, charbroiled pork and chicken.

If you're looking to explore Eurasian outside of the sandwiches, though, you're taking your palate into your own hands.

Need some mood music to get the appetite (and the heart) going? Give **Jazzed Café & Vinoteca** (8615 W. Sahara Ave. at Durango, 702-233-2859) a whirl. Considering that the restaurant occupies a modern glass-walled storefront (albeit with views on three sides), the atmosphere is rich, warm, sexy and rather mysterious. Paintings and art objects blend with wall and floor murals to create a sort of ragtag Gypsy chic—just the right feeling for some unique renditions of little-known but classic Italian dishes. But the real reason for going to Jazzed is to indulge in one of the risottos. Jazzed offers at least eight at any given time (including an astonishing one with anchovy and orange). Try the panna cotta, a velvety custard, for dessert. Also, there's always a pretty good (if a little New Agey) jazz band playing some smooth sounds to get you in the mood for food—and maybe a little love.

The Strip

Mr. Lucky's (4455 Paradise Rd. just north of Harmon inside the Hard Rock, 702-693-5000) is a straight-up diner that, despite being located inside a hotel-casino, has straight-up diner prices. And the food is pretty damn good. But the real reason to eat here—aside from the tuna melt—is the chance of spotting a real live rock star before and after a show at the neighboring music venue, the Joint. Billy Corgan, Dave Grohl, Lemmy Kilmister—they've all been spotted here at one time or another chewing on a cheeseburger and fries just like the rest of us (unless they're vegan or vegetarian, in which case it was probably a tofu burger). Breakfast is even better here than lunch or dinner. But if you see Dennis Rodman at a booth, please just leave him alone—he really, really doesn't enjoy signing autographs.

Another great affordable diner on the Strip is the **Peppermill** (2985 Las Vegas Blvd. S. just north of E. Desert Inn Rd. across from the Frontier, 702-735-4177), but this joint is even sexier than Mr. Lucky's. Nearing its 30th birthday, the Peppermill Fireside Lounge has long been a treasure hidden on the Strip. Commonly the backdrop for romantic types, the lounge helps couples celebrate occasions from honeymoons to 50th wedding anniversaries. But this 24-hour spot dubbed the Fireside Lounge isn't solely for lovers; it's a mellow place to begin or end a night on the city, especially crowded around the flaming water pit. The Fireside maintains large crowds, who usually don't end Friday and Saturday nights until 6 a.m. And the cocktail waitresses? These women definitely turn heads in their long, elegant evening gowns, serving Scorpions (the Fireside's signature drink). It's 60 ounces made up of six shots, fruit juice and ice cream for $10. Most drinks at the lounge are priced at less than $5.

Canter's Delicatessen (inside Treasure Island, 3300 Las Vegas Blvd. S. at Buccaneer, 702-894-7111) is pretty damn swanky, with Jetsons-type diner and stools and retro space-age tables and chairs. Indeed, you'll feel like a James Bond flying first-class en route to saving the planet. First, though, you'll need to eat some Jewish fare: potato knish, pastrami and corned beef, eggs, sour pickles, Hebrew Nationals, matzo ball soup. Most items on the menu run anywhere from $6.25

to $10.50. At these prices, you can afford to take in the free "Sirens of TI" show. (See the "Free Vegas" chapter, p. 57.)

Various Locations

Blueberry Hill diners are all over the valley, and they're great greasy spoons for breakfast, lunch and dinner. This '50s-style chain, for reasons beyond our understanding, doesn't get a lot of credit, even though it (at least in Vegas) serves some great Mexican *desayuno* (that's Spanish for "breakfast") with rice and beans, tortilla and chorizo (sausage). The burgers are fantastic, too. The fries are frozen, sure, but most everything else will satisfy at a solidly cheap price.

Capriotti's will make you wish you'd never contaminated your stomach with Subway, Quizno's, Schlosky's or any of those "healthier alternative sandwich shops." This Delaware-transplanted chain is so good precisely because it's so bad for you. Take, for instance, the "Bobbie," a Thanksgiving Day delight of turkey, stuffing and cranberry sauce. There's the "Slaw Be Joe," with roast beef, cole slaw, provolone cheese and Russian dressing. If you're a vegan or hanging out with vegans, don't worry. There are also plenty of meatless soy substitutes. Yeah, a large sandwich for $11 sounds pricey—except that a small is nine inches, a medium is a foot long and the large will take you two days to eat! Check out www.capriottis.com.

If you're from California, skip this paragraph, because you already know about the fantastic fast-food chain known as **In-N-Out** that cuts your fries to order using real, fresh potatoes. There's a reason the author of *Fast Food Nation* continues to eat here: The burger patties taste like real meat because, well, they *are* real meat—not that fake McDonald's gruel we all learned about in Eric Schlosser's book and Morgan Spurlock's documentary film *Super Size Me*. In fact, nothing is frozen at In-N-Out. Not the buns, the lettuce, the onions—it's all fresh. For less than five bucks, you get a Double-Double (without cheese), fries and a Coke. If you wanna splurge, get the cheese and ask for your Double-Double "animal style," which means they sautée the onions for you. Don't let the biblical passage ("John 3:16") at the bottom of your soft-drink cup scare you. This is a genuinely Christian family-owned chain that pays their lowest employee eight bucks an hour. Hell, these burger flippers make more than you do!

Fatburger isn't quite as good as In-N-Out for the simple reason that they serve frozen fries that don't taste any better than what you can buy at the store. But damn, their burgers are just as good if not better. Their patties are thicker than In-N-Out's, and unlike the Christburger chain, Fatburger serves some lethally good onion rings. And that relish they serve with the burgers? Gawd! The best way to work these joints is go in and say you left your buy-one-get-free coupon at home—you know, the one they gave you last time you were there? Most likely, the kids behind the counter will believe you. (They always buy my story, anyway.) Beware the Internet jukeboxes!

Tacos Mexico offers what are probably the tiniest, tastiest, 75-cent tacos in town. Again, if you're from Cali, no biggie here, as this is a L.A.-based chain. The best thing you could possibly do in Vegas is get a half-dozen tacos asado to go, grab a case of Tecate, head back to your sin bunker and just chill. But if you're feeling adventurous, you can actually dine in and experiment by loading your tacos, quesadillas and flautas with beef brains, beef cheek and pork maw. Oh, and lengua (tongue). And for a weird 99-cent dessert, don't forget to try a "cherry churro" or a "choco-flan."

A Note on Buffets

They're overrated. And the cheaper the price tag, the greasier and crappier the food turns out to be. If you want to stuff yourself with dry-as-a-bone fried chicken, freedom fries, and endless refills of Coke products like every other bloated tourist, go right ahead. Any of the hotel-casino buffet will suffice.

But if you've got $30 to burn, go to the **Buffet at Bellagio** (3600 Las Vegas Blvd. S., 702-693-7111). It's got all the fancy fixin's: grilled lamb, Kobe steaks, venison, duck, clams, sashimi, smoked sturgeon. The Tuscan décor is elegant, and uplifting, and the perfect antidote to a grumpy mood.

Another less sophisticated but no less impressive food arsenal is the **Buffet at the Golden Nugget** (129 Fremont St. at Bridger, 702-385-7111), which consists of carved roast beef, chicken, ham, turkey, endless salads and soups and homemade desserts. It's more "American"-centered than the Bellagio buffet, of course, but still incredibly satisfying.

Otherwise, **Lotus of Siam** (935 E. Sahara Ave., inside Commercial Center, 702-735-3039) has a delicious Thai lunch buffet from 11 a.m. to 2 p.m. during the week. It's the best thing you'll eat in Vegas. Trust me, okay?

CAFFEINATED SIN

It's tough being a literary animal in Las Vegas. There's only one independent coffeehouse left standing these days: **Brewed Awakenings** (2305 Sahara Ave. #F at Eastern, 702-457-7050), where you'll find the best lattes and mochas in town. Bagels, danishes and muffins are available, and there's original artwork, crafts and jewelry for sale, too. Sadly, there are no readings or literary events here. But it's still a wonderful place to read a book, especially if the weather's nice and there's a patio table open.

Other than that, there are a couple of decent chains. The first is **Coffee Bean & Tea Leaf**, which has several locations throughout the valley, but its foremost spot is across from UNLV (4550 S. Maryland Pkwy. at Harmon, 702-944-5029). There you'll find professors and madmen getting their caffeine fix and chatting about lofty topics. Don't get too excited, though. This place is a long way from Berkeley.

Another good place to sip cappuccinos and wear your black beret and turtleneck (in winter) is **Jitters Gourmet Coffee**, again with various locations in Las Vegas and Henderson. The best one is at 2191 E. Tropicana Ave. (at Eastern, 702-214-0888), simply because it's next to a great Indian lunch buffet called

Tamana (2295 E. Tropicana Ave., 702-798-9595).

And yes, I know **Krispy Kreme** (various locations) doughnuts are inappropriate for Atkins' dieters, but they're tasty and the coffee served with them is pretty good. More importantly, though, Krispy Kreme is leading the way in the Las Vegas Valley for free Wi-Fi (wireless Internet access).

If you prefer tea (for chrissakes, why? maybe for medical reasons), there's the wonderful landscape of **Tea Planet** (4355 Spring Mountain Rd. at Arville, 702-889-9989), which boasts a stylish Taiwanese décor and a seemingly infinite tea selection. There are also a few tasty snacks on the menu like the pickled ground pork and the red bean and rice ball soup.

HEALTH AND WELL-BEING

Las Vegas is all about feeling good. If you've overdone it on the "feeling-so-good-I'm-feeling-like-crap the next day" program, you've got plenty of options, whether that means a peaceful trip to a relaxing spa, the wholesome health food store for some organic veggies, or out of the city into the mountains for some fresh air.

Food Stores

Rainbow's End Natural Foods (1100 E. Sahara Ave. at Maryland, 702-737-7282) Fixin' for falafel? Need to stock up on soy flour? You'll find virtually every wholesome choice under the sun stocked on the shelves at Rainbow's End, and they've got a café to boot.

Whole Foods (8855 W. Charleston Blvd. at Durango, 702-254-8655) In addition to a creative, organic deli counter, the store offers wine and cheese tasting classes, cooking classes and yoga on the patio. Make sure you ask for an update on Whole Foods' most recent class schedule.

Healthy Eatin'

Café Sensations (4350 E. Sunset Rd. at Green Valley Blvd. 702-456-7803) Hours: 7 a.m. - 5 p.m. Healthy food never tasted so good. Sensations' mouth-watering offerings, chock-full of leafy goodness, makes eating salad sound like something you want to do. Salads start at around $7.

Go Raw (3620 E. Flamingo Rd. at Sandhill, 702-450-9007, www.gorawcafe.com) A "living cuisine and juice bar." Raw, organic, veggie. Prices start at $3.50 for aslice of living bread topped w/ "cheese", diced tomatoes, basil, garlic oil, pine nuts & hemp seeds. Hours: 10 a.m. - 9 p.m. M-Sat, 11 a.m. - 7 p.m. Sun.

Spas, Etc.

Ambiance Massage and Facial Day Spa (923 S. Rainbow Blvd. at Charleston 702-877-1144) With endless choices that include acupressure, detoxifying facials and citrus body wraps, Ambience goes beyond the typical day spa offerings. You don't even have to worry about getting there; chaffeur service is provided. Massage: 1 hr/$55; facial 1 hr /$60; citrus scrub $30; salt glow $25. Call ahead for appointment.

Canyon Ranch SpaClub (At the Venetian, 3355 Las Vegas Blvd. S. at Spring Mountain, 702-414-3600) Hours: 5:30 a.m. - 10 p.m. everyday. Vegas's premier spa is like a resort within a resort. It's got its own wellness center with dieticians, sports therapists and the like, work-out facilities with rock-climbing wall, a salon, spa and café. You could be in here for days. Massages, etc. are pricey but $35 gets you an all-day pass for use of every facility, including steam room, jacuzzi, and more.

Dolphin Court Salon & Day Spa (7581 W. Lake Mead Blvd. at Buffalo, 702-946-6000) Hours: 9 a.m. - 8 p.m. Mon.-Sat., Sun. closed at 6 p.m. Live the life of luxury, if only for a day. Pamper your tootsies, your coif and everything in between at this local favorite. $80/55 minute Swedish massage, $76/55 minute facial.

Nevada School of Massage Therapy (2381 E. Windmill Lane at Eastern, 702-456-4325) Student masseuses mean discounted massages. The joys of higher learning! Sat. and Sun. only, 8 a.m. - 5:30 p.m., walk-ins only, $32/hour Swedish massage.

Fresh Air
Red Rock Canyon National Conservation Area (1000 Scenic Drive at W. Charleston, 702-515-5350, www.red rockcanyon.blm.gov) Located 17 miles west of the Strip on Charleston Boulevard, aka Highway 159. Miles and miles of nature trails don't leave much standing between you and the bighorn sheep, deer and other inhabitants you'll find at the base of the Spring Mountains. Early mornings are your best bet to encounter such creatures. Entry fee: $5/vehicle, entrance to 13-mile scenic drive. Campsites: $10/day. Drinking water and restrooms provided. There's no shade. Cash only.

WILD WEDDINGS

Sure, you could always get married by Elvis, but that's just *so* 1996. Try these truly one-of-a-kind weddings—you'll have to have pictures to make your friends believe it.

Gothic: At **Viva Las Vegas wedding chapel** (1205 Las Vegas Blvd. S. at Charleston, 702-384-0771, www.vivalasvegasweddings.com). Step out of the coffin and into the fangs of your beloved. $150+.

Lancelot: At **Excalibur's Canterbury wedding chapel**, (3850 Las Vegas Blvd. S. at Tropicana, 702-597-7278, www.excalibur.com/weddings). Your love for Renaissance faires (and each other) has gone wild. $365+. Medieval wedding costumes extra.

Pirate: At **Treasure Island** (3300 Las Vegas Blvd. S. at Spring Mountain, 702-894-7444, www.treasure island.com/pages/amenities_weddings). Walk the plank or take the plunge—you decide. $500+. Pirate costumes extra.

Star Trek: At the Las Vegas **Hilton** (3000 Paradise Rd. at Riviera Blvd., 702-697-8700, www.startrek

exp.com). You'll be beaming when you get married on the hull of the USS Enterprise, the Klingons, Borgs, and Ferengi at your side. $500+. Bring your own costume.

Madame Tussaud's: At the **Venetian** (3377 Las Vegas Blvd. S. at Spring Mountain, 702-862-7805, www.venetian.com/attractions/madame_tussauds.cfm). The bride won't be the only one blushing, especially when guests get too close to J.Lo. $3500+

Railroad: At **Nevada State Railroad Museum**, (600 Yucca St., Boulder City, at U.S. 93, 702-486-5933). Revel in the treatment railroad magnates were used to as the defunct Boulder City Railroad takes the tracks just for you.

KID-FRIENDLY STUFF

Although the Visitor & Convention Bureau may try to persuade you otherwise, there are more than a few places kids can be kids in Las Vegas. At most of them, big kids will want to play, too. A decent website with more kid-friendly info is **www.lasvegaskids.net**.

Black Mountain Recreation Center Aquatic Complex (599 Greenway Rd. at Horizon Drive, Henderson, 702-267-4070) Open summer only between Memorial and Labor Days. Hours: Mon-Sun 11 a.m.-6 p.m. Hit the water in Henderson at this aquatic center with a three-loop water slide and teardrop waterfall. There's also a game room for those who left their Gameboys behind. Fee: around $3.

Bonnie Springs Old Nevada (1 Gunfighter Lane at Charleston, 702-875-4191, www.bonniesprings.com) Hours: 10 a.m.-5 p.m. The Wild West is alive at this 115-acre ranch located 30 minutes from the Strip, where you'll find all the obligatory trappings of the 1800s— stagecoaches, tumbleweeds, a shoot 'em up gun battle, and a hangman swaying in the wind. Horseback trail rides available ($30/hr). Fee: $7/car weekdays, $10/car weekends.

Coney Island Emporium (At **New York-New York**, 3790 Las Vegas Blvd. S. at Tropicana, 702-736-4100, www.coneyislandemporium.com) Just as big as Gameworks, but with the feel of a real midway. You can even have a Coney Island dog at Nathan's Famous Hot Dogs—just wait till after riding the Manhattan Express coaster.

Desert Breeze Skate Park (8275 Spring Mountain Rd. at Durango, 702-455-8200) Make like Tony Hawk at Nevada's largest skate park. Boarders, bladers and bikers are all welcome on the concrete course, which is intermediate with some beginning elements. Free.

Event Center Las Vegas (121 E. Sunset Rd. at Las Vegas Blvd., 702-317-7777) Hours: W-F 3-9p.m., Sat and Sun, Noon-9p.m. Wish you could hit one out of the park at Wrigley Field? Up-and-coming sluggers can hone their craft at this sports park, which also features a NASCAR go-kart track. Prices: $2+

GameWorks (3785 Las Vegas Blvd. S. at Tropic-ana Ave., 702-432-4263) The ultimate video playground for kids, thanks to Steven Spielberg. The basement space has more than 250 games to choose from, so by the time you've had your fill, you'll look like a mole getting reaccustomed to the light.

Jillian's at Neonopolis (450 Fremont St. at Las Vegas Blvd., 702-759-0450) Hours: 11 a.m.-9:30 p.m. Two levels of skill games, video games, bowling and food. Plus, they have air hockey!

Las Vegas Cyber Speedway (At the Sahara, 2535 Las Vegas Blvd. S. at Sahara 702-737-2750) Hours: Sun-Th 12-7 p.m., F-Sat, 11 a.m.-10 p.m. Make your buddies eat dust as they try to catch you in this track modeled after the Las Vegas Motor Speedway. You can customize your simulator for maximum torque and aerodynamics, or just blast the 15-speaker sound system. Riders must be 54 inches tall. $10.

Las Vegas Mini-Gran Prix (1401 N. Rainbow Blvd. at Vegas Dr., 702-259-7000,www.lvmgp.com) Hours: Sun-Th 10 a.m.-10 p.m., F-Sat -11p.m. Finally, something the Northwest can be proud of! You won't be relegated to go-karts here (unless you choose one yourself). In the banked sprint track you can really pick up speed. Tickets: $5.50

Rock-climbing wall at Galyan's (At the Galleria mall, 1300 W. Sunset Rd. at Stephanie, 702-258-1381) The most fun you can have at a sporting goods store. That means you don't have to bring your own gear. Free.

Shark Reef (At Mandalay Bay, 3950 Las Vegas Blvd. S. at Mandalay Bay Dr., 702-632-4555, www.man dalaybay.com/entertainment/shark/) Hours: Daily 10 a.m.-11 p.m. There's just something cool about fish so large they could snap your femur. The carnivorous water monitor, piranha and golden crocodile also inhabit the exhibit. Keep your hands to yourself! Admission: Adults $15.95, Children 12 years old and younger $9.95, Children 4 and younger admitted free.

TREASURE HUNTING

Make a statement: Pay as little for your fashion as humanly possible. Whether you're looking for goth gear, décor for the home, or dress-up clothing, these shops will reward your miserly side if you're willing to dig a little. Another man's treasure indeed. Call ahead to make sure they're open when you're ready for a few bargains.

Attic Rag Co. (1018 S. Main St. at Charleston Blvd., 702-388-4088). For the impossibly hip and those of us trying to be. It's cooler than you'll ever be: The Attic was featured in its very own Visa commercial. Movie props, furniture and funky fashions are just the beginning.

Big Lots (Various locations; 3430 E. Tropicana Ave. at Pecos Rd., 702-454-5055). Target for the monetarily disadvantaged. Score cheap linens for the kitchen, laundry soap and food items you thought were discontinued. Plus, socks and underpants for a buck!

Buffalo Exchange (4410 S. Maryland Parkway at Flamingo Rd., 702-791-3960). Boutique with previously owned fashions for skinny girls and guys. Stock up on your Bebe and Banana Republic.

Cash 4 Chaos (4110 S. Maryland Parkway at Flamingo Rd., in the same strip mall as Buffalo Exchange, 702-699-5617). Get your goth on with velvety capes, slasher videos and enough black to give Robert Smith a wet dream. Finally, you'll be able to replace that torn-up Siouxsie poster.

Charleston Outlet (1548 E. Charleston Blvd. at 15th St., 702-388-1446). Bargains galore if you're willing to dig a little. Ladies fashion alone takes up its own room. Inventory changes daily, and there are always some dynamite discards.

D'Loe House of Style (220 E. Charleston Blvd. at Casino Center Drive, 702-382-5688). Dress up like gram and gramps! Vintage clothes, shoes and accessories from the 1930s to 1970s. Who said pea coats were out of style?

Deseret Industries Thrift Store (1300 Las Vegas Blvd. N. at Owens Ave., 702-649-8191). Warehouse-style thrift store with everything under the sun. Drop by if you can get past all the bums stumblin' down the north boulevard.

Goodwill of Southern Nevada (Various locations; 4632 S. Maryland Parkway at Harmon Blvd., 702-895-9917). Begs the question, "Why do people throw this stuff away?" Nice collection of used T-shirts, stirrup pants and athletic wear. You may even find a nice punch bowl.

Home Consignment Center (2360 E. Serene Ave. at Eastern Ave., 702-616-0989). Most of this stuff looks like it came right out of the showroom—and the wood isn't even painted white! Hard to find a matching set of barstools, though.

Opportunity Village Thrift Store (921 S. Main St. at Gass Ave., 702-383-1082). The dumping grounds for furniture donated to raise money for the disabled. Go buy a lamp and help disadvantaged adults.

The Refinery Celebrity Resale Boutique (3827 E. Charleston Blvd., at Sandhill Rd., 702-384-7340) When a Gucci has outlived its life with the moneyed set, it ends up here. Give a pair of Kate Spade shoes a better life.

Ritzy Rags (2550 S. Rainbow Blvd. at Sahara Ave., 702-257-2283). Although the term "ritzy" should be abhorred, this consignment shop has one of the best designer inventories. Snatches up items from estates, so you could end up wearing some dead guy's suit.

Salvation Army Family Superstore (360 N. Stephanie St. at Warm Springs, 702-436-3100). One of the few without clothes packed together like a block of ice and merchandise strewn everywhere. Roomy dressing rooms, with (gasp!) full-length mirrors.

Savers (Various locations; 1100 E. Charleston Blvd. at Maryland Parkway, 702-474-4773). A garage sale moved indoors, but with discount days and special sale racks galore. Guess it goes with the territory when you're a not-for-profit chain.

If you simply must hit a flea market, check out **Broadacres Swap Meet** (2930 Las Vegas Blvd. N., North Las Vegas, 702- 642-3777). Fridays, Saturdays, and Sundays 6:30 a.m. till dusk. 400 vendors, new and used. $1 admission includes parking.

FREE VEGAS

Most of Vegas's big Strip casinos and stores offer some kind of free attraction to lure in the visitors. Now that you know better, check out these freebies while holding onto your purse strings. Call ahead for hours.

Bellagio Fountains (3600 Las Vegas Blvd. S. at Flamingo Rd., 702-693-7111). The fountains snake and twist like a hooker who just found herself a footlong.

Bird Show at the Tropicana (3801 Las Vegas Blvd. S. at Tropicana Blvd. 702-739-2411) They like to ride their bicycles. And sweep so low they almost hit old ladies with beehives in the head.

Ethel M Chocolates Factory & Cactus Garden (Two Cactus Garden Drive, Henderson, at Sunset and Mountain Vista, 702-433-2500) Nothing perks you up like free chocolate.

Flamingo Wildlife Habitat (3555 Las Vegas Blvd. S. at Flamingo 702-733-3111) Bet Bugsy would have never dreamed penguins would be pooping on his lawn.

Fountain show at the Forum Shops (3570 Las Vegas Blvd. S. at Flamingo Rd. 702-731-7110) Roman

God of thunder battles a wicked beast. Or is that the Cheesecake Factory?

M&M World (3785 Las Vegas Blvd. S. at Tropicana Ave., 702-736-7611) Anything you could possibly dream of with M&M's emblazoned on it. Yes, adults actually do wander in here.

Masquerade Show in the Sky at the Rio (3700 W. Flamingo Rd., 702-777-7777) See all those people walking around with plastic beads? In here, it's cool, at least till the show's over.

Tiger Habitat at the Mirage (3400 Las Vegas Blvd. S., 702-791-7111) Siegfried & Roy spent millions so you could watch the tigers paddle around in water cleaner than what's in your bathtub.

The Sirens of TI at Treasure Island (3300 Las Vegas Blvd. S., 800-288-7206) Scantily-dressed sirens battle swashbuckling pirates. Aaaargh! Yeah, right.

WEIRD VEGAS CONVENTIONS

Three sex industry conventions take place every January in Las Vegas. The first is **Internext** (www.internext-expo.com), followed by the mammoth **Adult Entertainment Expo** (http://show.adultentertainmentexpo.com), both organized by the trade publication *American Video News* (AVN). **Bondcon** (www.bondcon.com), a BDSM/fetish convention, is an offshoot of **AEE** and runs concurrently at a different location.

The first half of the **Adult Entertainment Expo**'s opening day is limited to the trade, and is fairly reserved, but the second half is for the fans with more porn stars and video screens showing explicit scenes. Flynt might be signing copies of *Hustler* at his company's exhibit. Ron Jeremy might be lurking about, and 50 Cent, Hulk Hogan and Mike Tyson could cause minor commotions when they show up.

Forget porn. What if what you really need to cover all that skin is a suit of armor? Or a handmade French sword from the 18th century? Or a pirate's cannonball? All this and more can be yours to buy, sell, trade and appraise at the **Las Vegas Antique Arms Show** (www.antiquearmsshow.com), which is considered by many to be the best collectable antique arms exhibit on the planet.

Forget rusty old scabbards. What about your knack for impersonating a celebrity? If you've got what it takes to imitate Britney Spears, for instance, you can get all "Toxic" at the **Celebrity Impersonators Convention** (www.celebrityimpersonators.com, 1-888-771-6611), featuring celebrity look-alikes and female impostors. Just think of it! An army of Chers, Madonnas and Celines fighting it out in Vegas like they should be. Yes, the stars are in perfect alignment tonight.

Every year, the **Far West Popular Culture Association** (www.unlv.edu/ colleges/liberal_arts/english/ popcul/pcr.htm) meets at the University of Nevada, Las Vegas, in January. Sponsored by the English Dept., this gathering involves stuffy academics getting their geek on with topics like Buffy the Vampire Slayer and Harry Potter—y'know, pop culture. Put on your thinking cap and fire up that GameBoy: It's academic time!

And after gorging on junk culture, you can stuff yourself silly at the **International Pizza Expo** (www.pizzaexpo.com), the nation's oldest and largest pizza tradeshow in the world, where leading Italian pie-makers get together for seminars with titles like "Pizza Crust Boot Camp." Man, if only we all owned pizza joints!

But hell, we Americans are always deemed dummies simply because of our insatiable appetite for fast food and disposable culture. Well, most likely you'll encounter some real dummies at the **Vegas Ventriloquist Festival** (www.inquista.com). But you'll also learn how to throw your voice. Maybe.

Other great conventions that take place in Vegas are the **International Swizzle Stick Collectors Association**, and **G2E: Global Gaming Expo and the**

Amusement and Music Operators Association. For a complete calendar, visit the **Las Vegas Convention and Visitor Authority** website at www.lvcva.com.

TECHNO VEGAS

Tech Conventions

Comdex (www.comdex.com) is not the largest technology conference in the world. That distinction belongs to the annual CeBit convention held in Hanover, Germany. But Comdex, with a past average of 225,000 visitors each November, is certainly the largest technology convention in the United States.

Comdex began as the Computer Dealers Exposition in 1979, with an attendance of about 4,000 people—enough to fill only a single ballroom. But a decade later, it was the most important computer convention around, filling up the Sands Expo Center and causing endless headaches for locals. Traffic reaches epidemic proportions and local nightclubs close down for private industry parties where nerds mill around, get drunk and discuss the state of the technology industry, which usually changes right after Bill Gates gives the customary opening-night keynote speech.

It's a weird thing to watch the nerd world interact with the shimmering decadence of Las Vegas. For many of the geeks, it's their one chance each year to cut loose—which usually involves getting badly drunk, throwing massive amounts of money away on the slots, and having creepy convention sex with the armies of bimbos—both male and female—whom exhibitors hire to show off their wares. Old Comdex veterans act like jaded Vegas locals, nonchalantly giving newbies advice on which strip clubs and buffets to hit at night. During the day, they wander the floor of the Las Vegas Convention Center, looking at all the new business-to-business solutions and waiting for night to fall. I often suspect that many of the worst business decisions that occurred during the dot.com era were made by Comdex visitors, their brains foggy and hung over.

In recent years, Comdex has waned along with the fortunes of the technology industry. The permanent demise of Comdex would have serious repercussions not only for the technology industry, but for Las Vegas's economy. After all, the convention generates an estimated $250 million dollars per year in revenue for the city (according to the Las Vegas Convention and Visitors Authority). And for local businesses that rely on Comdex to be the biggest convention of the year—like GES Exposition Services, which handles local construction and services for the show—such a loss would be disastrous. The 2004 convention was "postponed" but there one is currently scheduled for November 2005.

The **Consumer Electronics Show** (www.ces web.org), on the other hand, is only getting bigger. The 2003 show was the largest yet according to the CES website, with 116,687 visitors and 2,283 exhibitors. Since 1970, CES has showcased the latest in consumer electronics—everything from car stereos to home theater systems and ... er, personal massagers.

CES is a profoundly boring convention, though, unless you happen to be the sort of person who gets aroused looking at the latest in speaker technology.

DefCon (www.defcon.org) is perhaps the most interesting technology convention held in Vegas. Each August, several hundred geeks converge on the Alexis Park Hotel for the convention, which ostensibly deals with computer security—but is really an excuse for extremely intelligent, socially maladjusted hackers to get together, drink beer, get laid, and talk about new ways to stick it to the Man.

At the end of the convention, all the hackers head back to the Bay Area or Los Angeles or Seattle or New York, leaving nothing in Las Vegas except for the vague feeling that we might be a little behind the times.

Websites

Las Vegas does not love the Web. For a major metropolitan area, Vegas may feature the smallest number of local websites. Because Vegas has traditionally shied away from courting the technology industry, there aren't a lot of tech-savvy people out here. Consequently, Las Vegas is short on startups, weblogs, community sites ... pretty much the anchors of most Internet communities.

Except for porn, of course. Vegas is one of the two centers of Internet porn: the other is the Netherlands, but they handle all the stuff we'd get arrested for.

"House of Cards" by Mike Davis (www.rut.com/mdavis/housecards.html) Urban theorist Mike Davis is most well-known for *City of Quartz*, his deconstruction/meditation on the secret history of Los Angeles. In this essay—written for *Sierra* magazine in 1995—Davis opens a can of whup-ass on our fair city.

For example: "The Las Vegas 'miracle,' in other words, demonstrates the fanatical persistence of an environmentally and socially bankrupt system of human settlement, and confirms desert rat Edward Abbey's worst nightmares about the emergence of an apocalyptic urbanism in the Southwest. Although postmodern philosophers (who don't have to live there) delight in the Strip's supposed 'hyperreality,' most of Clark County is stamped from a monotonously real and familiar mold. Las Vegas, in essence, is a hyperbolic Los Angeles, the land of sunshine on fast forward."

Oh, yeah.

www.YourLocalScene.com Founded in 1998 by local scenester Jeff Higginbotham, yourlocalscene.com is both a Vegas music portal and a forum for artists to get their work online. Featuring local talent like singer/songwriter Michael Soli and New Wave/garage rockers the Killers, YourLocalScene.com is one of the best places to dive into the Vegas music world.

www.VegasGoths.com Surprisingly enough, for a place as relentlessly sunny and banal as Vegas, there's a fairly large goth subculture here. Correction: There are a lot of fat teenagers in bondage collars and Cradle of Filth T-shirts who would *love* to be goth, but aren't very convincing.

Separate the Peter Murphy from the chaff at www.vegasgoths.com, a forum/event calendar for the Vegas goth scene. Discuss all things dark and dreary, including the possibility that anybody will ever get Nick Cave to play the Huntridge Theatre. And hookup for a game of "Goth Kickball," which is truly frightening.

www.KeanuVision.com Keanuvision (run by a young lady from Las Vegas who goes by the handle of "Krix") is the ultimate resource for Keanu Reeves fans. In addition to finding out all the latest Keanu news (is that "Kean-ews"?) via Krix's blog, you can download desktop icons, CD covers from Keanu's band, Dogstar, and objects for the *Sims*. The greatest thing about Keanuvision is the large soft-lit photo of Mr. Reeves wearing leather pants, a leather vest, and not much else. It is the mark of a truly dedicated fan (as is Krix's ability to actually listen to Dogstar).

LVACTS webcam (www.lasvegasnow.com/ lvactspopup.html) So postmodern, it's almost revolting, the Las Vegas Area Computer Traffic System is responsible for those weird surveillance feeds overlooking various intersections that you can find on local access TV early in the morning. For those of us who don't watch TV, KLAS has thoughtfully provided a webcam page, showing random snapshot images of various Vegas hotspots, such as the corner of Flamingo and Maryland, and City Hall. There may be an entire grad student dissertation here. The whole idea of living under constant surveillance is more than a bit creepy.

www.guggenheimlasvegas.org This may be the most well-designed website in Las Vegas. Pity it's almost devoid of information. The Guggenheim/ Hermitage Axis of Boredom may be the only real culture we have in Vegas, though, so you might as well check it out. And don't forget to stop into the online museum store and pick up a Roy Lichtenstein shirt, so you can be the envy of all the other parking lot attendants!

www.summerlin.com Summerlin? It's a major upscale suburb on the edge of Vegas that borders Red Rock Canyon. Going to Summerlin's corporate website is like bathing yourself in George Bush's AmeriKKKa. The first words that stick out on the main page are "master plan", a term I associate primarily with supervillains from Bond movies, any one of whom might have decided to build Vegas's own Triumph of the Swill burbclave as a lair for his evil SUV-driving yuppie henchmen. The website is full of ultra-freaky touches, like the "Spiritual" page, which explains that Summerlin is home to "a growing list of worship facilities." Summerlin is also full of "medical facilities," "shopping facilities" and "chugging-a-latte-on-your-way-to-Pilates-class facilities." If this is truly the future of urbanism, Christ annihilate us. Visit this page and know your enemy.

www.zenarchery.com Zenarchery.com is perhaps the greatest website to ever be designed and developed in Las Vegas. Featuring the words of America's best-loved commentator and deranged futurist, Dr. Joshua Ellis, Zenarchery has been proven in scientific tests to increase your I.Q., raise your libido and shrink your waistline. Plus, you can download music and fly wallpaper. Not to mention reading the Bad Doctor's everyday thoughts, such as "America is a nation of dogs" and "With the greatest sincerity I can muster, may you motherfuckers rot in hell." Fun for the whole family!

TECHNOMADISM

What is life without digital entertainment, devices, toys, and other useless shit? If you're looking for the best free Wi-Fi hotspots (that provide wireless Internet access if you're using your laptop), LANs, arcades or places where you can purchase replacement parts for your electronic devices, you've turned to the right page.

Free Wi-Fi Hotspots

Oddly (or appropriately) enough, **Krispy Kreme** is leading the charge in providing free Wi-Fi (wireless Internet access) in the Las Vegas Valley. So we recommend getting some doughnuts and coffee while you surf the Web. You can find a Krispy Kreme at: 1331 W. Craig Rd. at Martin Luther King, 702-657-2527; 9791 S. Eastern Ave. at Silverado, 702-617-4529; 7015 W. Spring Mountain Rd. at Rainbow, 702-222-2320.

Coffee Bean & Tea Leaf 4550 S. Maryland Pkwy. Ste. A, at Harmon, 702-944-5029; 7291 W. Lake Mead Dr. at Tenaya, 702-944-0030.

East Boy Japanese Café 4755 S. Maryland Pkwy. just north of Tropicana, 702-798-1777.

Apple Store (and nearby area outside store) at Fashion Show Mall 3200 Las Vegas Blvd. S. Ste. 1760 at Spring Mountain, 702-650-9550.

Clark County Library 1401 E. Flamingo Rd. one block east of Maryland Pkwy., 702-507-3400.

Panera Bread 605 Mall Ring Circle at Sunset and Galleria across the street from Sunset Station (in Henderson, Nev.), 702-434-4002.

Buffalo Wild Wings (617 Mall Ring Circle at Sunset and Galleria across the street from Sunset Station in Henderson) 702-456-1237.

And if you want to stoop so low as to actually pay money for Wi-Fi, you can set up laptop shop at certain **Starbucks**, **Borders Books & Music** or **Jitters Gourmet Coffee** locations.

Hotels that offer complimentary wireless Internet: **Residence Inn** (370 Hughes Center Dr. at Flamingo and Paradise, 702-650-0040) and **Fairfield Inn** (3850 Paradise Rd. at Flamingo, 702-791-0899). Otherwise, it's $10 a day for Wi-Fi at **Four Seasons** (3960 Las Vegas Blvd. S. at Russell, 702-632-5000), the **Rio** (3700 W. Flamingo Rd. at Valley View, 702-777-7777), **MGM Grand** (3799 Las Vegas Blvd. S. at Tropicana, 702-891-1111) and the **Mirage** (3400 Las Vegas Blvd. S. between Spring Mountain and Flamingo, 702-791-7111).

LANs and Cybercafés

The last couple of years have seen an insane amount of cybercafés sprouting up around the valley. These places are not just for gamers, though; you can check web-based e-mail accounts like Yahoo!, browse the Internet (presumably for porn) and do other, cyber-geek-related activities at these places. The first three are within one block of each other (all across the street from

UNLV) and are widely regarded as being the best in town. **Cyberzone** is actually world-famous, and is visited randomly by some of the best gamers in the world. Most are open well past midnight and will only close if there isn't a crowd; call for late-night availability. Prices (off-Strip) are universally cheap ($2-3 per hour):

University District

Cyberzone 4440 S. Maryland Parkway at Harmon, across from UNLV, 702-732-2249

Planet PC 4550 S. Maryland Parkway at Harmon, across from UNLV, 702-262-7820

PC Intercrew 4632 S. Maryland Parkway, 702-262-6788

Strip

CyberStop Internet Café 3743 Las Vegas Blvd. S., #112B, in the Hawaiian Marketplace next to the 7-Eleven inside the gift shop, 702-736-4782 ($12/hour due to its Strip location, closes at 2 a.m.)

Central

GameZone 4011 W. Sahara Ave. at Valley View, 702-252-0755

Cyberpunx 4620 Meadows Lane at Decatur, 702-88O-PUNX

Traditional Arcades

Unfortunately, the rise and proliferation of cybercafés has utterly annihilated nearly all traditional arcades operating off of the Strip. Therefore, your best bets are the major hotels and casinos. The **Luxor** has one of the biggest and best arcades that's not yet prohibitively expensive and has a few classic machines.

Of course, there's always **GameWorks** (3785 Las Vegas Blvd. S. at Tropicana, 702-432-4263), the casino-sized, super-premium arcade on the Strip. Head there only if you're willing to shell out some serious cash; the games are expensive.

Where To Buy Stuff

There are a bunch of places to go if you find yourself in need of technological accessories (i.e., forgot to bring a USB cable, batteries, memory sticks, etc.). In a pinch, you can always head to an office supply store like **Office Max** or **Office Depot** (numerous locations throughout the valley). Places like these are especially good if travel time to one of the electronics stores is too extreme. Here are a few convenient places:

University District

Best Buy 3820 S. Maryland Pkwy. (just north of Flamingo) 702-732-8342

Westside

Best Buy 2050 N. Rainbow Blvd. (at Lake Mead Blvd.) 702-631-4645; also (Summerlin) 10950 W. Charleston Blvd. (at I-215) 702-228-6492

Central

Circuit City 5055 W. Sahara Ave. (at Decatur) 702-367-9700

Eastside

Circuit City 4860 S. Eastern Ave. (at Tropicana) 702-898-0500

Strip

Fry's Electronics 6845 Las Vegas Blvd. S. (between Sunset and Warm Springs) 702-932-1400

For computer-specific needs, we recommend **Laboratory Computers** (3421 E. Tropicana Ave. at Pecos, 702-898-3700, www.laboratorycomputers.com). Besides, they have great ads featuring goth-punk chicks straddling hard drives.

CLICK

THE GREAT OUTDOORS

There are three incredible outdoor places on the outskirts of Vegas that are definitely worth visiting. Scratch that: They're all "must-sees." The first and most important is the **Red Rock Canyon** (702-515-5350, www.redrockcanyon.blm.gov), a 200,000-acre National Conservation Area that's just 20 minutes west of the Strip. Appreciating the desert's natural beauty might take a while for folks who live in lush, verdant areas of the country. But once you "get it," you'll be haunted by what you see. Indeed, the vivid-red rocks are like something an astronaut might encounter on Mars. There's a scenic, 13-mile loop that you can drive, but park the car at some point and hike one or more of the gorgeous trails (which are designated "easy," "medium" and "difficult" so that you don't end up food for vultures if you're not physically or mentally prepared for what lies ahead). Sagebrush, Joshua trees, waterfalls, red-tailed hawks, burros, salamanders—they're all here for you to enjoy in their natural habitat of the Mojave desert. Bring water and snacks to stay hydrated and energized.

Another—and even better—place to experience the natural beauty surrounding Las Vegas is at the **Valley of Fire State Park** (55 miles northeast of Vegas, take I-15 north to State Road 169 and get off at exit 15, 702-397-2088, http://parks.nv.gov/vf.htm). This is where you wanna pull a Jim Morrisson and just trip amid the already-psychedelic sandstone shapes and rock formations (with someone responsible to help out in case things go badly, of course). At dawn and dusk, everything takes on an otherworldly hue, making this one of the most beautiful places on the planet. In addition to water and a lunch, make sure and bring some stoner rock (Queens of the Stone Age, Monster Magnet, Verbena) and some drone-pop (My Bloody Valentine, Sci-Flyer, Spaceman 3) for the car stereo, as you maneuver across the eerie landscapes, including geologic formations like Silica Dome, Elephant Rock and Seven Sisters, which are song titles in and of themselves. Don't miss the mind-blowing 3,000-year-old petroglyphs.

If, for whatever reason, the desert's charms are lost on you, then the cool, lush pine forests of **Mount Charleston** (45 miles northwest of Vegas, take Highway 95 to State Road 156 up Kyle Canyon Road, 702-515-5400, www.fs.fed.us/r4/htnf), located inside the Toiyabe National Forest, will comfort and console. The campgrounds are open from May to October (unless there's, like, a fire or something). There are some fantastic trails here worth hiking, even if the "easy" ones are perhaps a little steeper and more arduous than those at Red Rock or Valley of Fire. If you don't have the time or inclination to pack sandwiches, then grab lunch or dinner at the **Mt. Charleston Lodge** (Kyle Canyon Rd., 702-872-5408, www.mtcharleston lodge.com). The appetizers include typical finger food selections like chicken fingers ($6.95), fried mozzarella ($4.95) and potato skins ($4.95). The pricey selections in the "entrée" section never get any more daring than chicken parmigiana ($17.95) or fettuccini alfredo, although the latter can be prepared plain ($11.95), with chicken and broccoli ($16.95) or with salmon and mushrooms ($17.95). The "From the Broiler" portion of the menu offers four types of steak ($21.95 to $27.95), while the seafood section (all choices $19.95) features salmon or tuna broiled with dill butter sauce, shrimp tempura, shrimp scampi or orange roughy sautéed or piccata style. Or try the **Canyon Dining Room & Cliffhanger Lounge at the Mount Charleston** Hotel (Kyle Canyon Rd., 702-872-5500, www.mtcharlestonhotel.com). For casual diners, they offer pizzas ($9 and up), burgers and sandwiches ($8.50 to $10.95) and finger foods like chicken fingers ($7.25) and nachos ($7.95). Entrées include six different pastas ($16.75 to $26), four steaks ($22 to $24) and three seafood dishes such as pan-fired mahi-mahi coated in a golden pecan crust and ginger beurre blanc sauce ($23). You can also order buffalo, elk or ostrich burgers ($11.25), or entrées of New Zealand elk chops with tarragon sauce, pheasant with a cognac honey pinenut sauce or ostrich with sundried cranberry bordelaise ($30 each or $45 for a platter of all three).

If you'd prefer to "go nuclear" in outdoor Nevada, you should take a tour of the **Nevada Test Site** (www.nv.doe.gov/nts, 702-295-0944), which is located 65 miles northwest of Las Vegas. The U.S. Department of Energy and the National Nuclear Security Administration Nevada Site Office only offer a tour once every month, so you need to register beforehand by phone, mail or e-mail. (Check the website for details.) What happens is this: Jump on a bus in North Las Vegas and drive to the site, where you're given a tour of the 1,375-square-mile facility. They don't provide lunch, so pack your own and dress for the crossing of rugged terrain—no shorts or sandals, please! And don't be pregnant, OK? You don't want your baby to grow up to be an X-Man or something. (Or do you?) Anyhow, the site is an eerie place and well worth a visit, especially if you're interested in America's nuclear history, or just like to fantasize about giant-monster movies like *THEM!* Don't plan on scrapbooking this part of your Vegas experience, though: cameras, video and tape-recording devices and even binoculars are all prohibited.

Before you wander the nuclear wastelands, you need to experience the **Atomic Testing Museum** (755 E. Flamingo Rd. at Swenson, 702-794-5151) on the UNLV campus. It's a brand-new museum, and one of the most fascinating places to spend an afternoon. Interactive features, pieces of old reactors and such serve to remind us just how serious the Cold War really was.

The gift shop is ultra-cool, with books and DVDs featuring images of atomic detonations in the Nevada desert. Check it out.

GOD DAM

In the hours after midnight, from the Arizona side, the **Hoover Dam** (30 miles southeast of Las Vegas on US Highway 93 at the Nevada-Arizona border, 702-294-3517, www.usbr.gov/lc/hooverdam/service) is as massive and quiet as the grave of some ancient and profoundly egotistical emperor.

The only noise is the chirping of crickets, the whipping of the wind and the occasional growl of a traveler's engine as he makes his way up the sinuous, narrow road towards Kingman, Scottsdale and Phoenix.

At night, it is a constellation of orange sodium streetlights, punctuated by the occasional halogen spotlight. Only the faintest light reaches the cliff faces and mountains around it, like ghosts half-glimpsed. Beyond their silhouette is the purple glow of Las Vegas, on the other side of Boulder City and Railroad Pass.

Like most of the places that people build purely for functional reasons, it is very lonely at night; nothing here is built to human scale. It reminds me of the freeways of Los Angeles at night; thousands of miles of concrete and empty space, silent save when some car rushes by, going from one place to another. Hang long enough at a truck stop in the middle of America somewhere, and you will get that same feeling of loneliness. You can stand on the edge of where the light is, and look out forever at the dark.

I believe that this is all that will be left when you and I and George W. Bush and Mick Jagger and Osama Bin Laden and William Shakespeare are long forgotten, an ice age or two down the road: just a trillion tons of concrete spread across the planet, laid out in shapes and patterns that, without the animus of civilization to drive them, will make no sense at all to whatever possible observer might come across them.

Or perhaps our media might somehow survive the millennium; perhaps our observer will look at his or her children and say, "This was once a desert; humans lived here, and changed the course of vast rivers to bring water, and therefore life, to this place."

"That's very silly," the children might respond. "Why didn't they just move someplace nicer?"

The Hoover Dam is a monumental structure, but there are still people alive who remember when it did not exist. Barring catastrophe, however, there is no one alive now who will see the Colorado River valley unencumbered by this great wall. If it were to fall, most of the cities of the southwestern United States would cease to exist in a matter of months without its electricity and its water. But the dam will still exist long after those cities have become archeological excavations. It was built well by the tens of thousands of unemployed men who laid its foundations in the dim days of the Great Depression. None of those men, contrary to popular legend, are actually buried within the dam, and though it resembles an imperial tomb, there are no unquiet ghosts here that I can see. If there were, they would face a hard eternity; the Hoover Dam will be here to haunt forever, or as close to it as we can possibly imagine.

A quarter mile away flies an American flag, its

shadow morphing across the rock face behind it. As I drove here, I was stopped by a policeman who inspected my van, presumably to ensure that I wasn't carrying a nuclear bomb or a bunch of Canadian immigrants to Arizona.

As opposed as I am to homeland security, I understand in this case. The Hoover Dam is an important object, not just for practical reasons but for patriotic ones. Outside of our great highway system and our railways, it is the biggest and most complex building project that our government has ever undertaken. It is a symbol of pride. It must be defended, because—like the Jefferson Memorial and Mt. Rushmore—it's as close to an official holy place as America has.

The hour groweth late; I think I'll leave you here, staring out across the black water shimmering under the streetlights. I'll smoke another cigarette, drink my power drink and drive back into the light and noise of the city. I'm glad I came here, though, finally. I will not forget it.

BIG HAIR AND
EARPLUGS

Vegas
Music
Scene

You can get a firm grasp on what's musically oc-
curring in Vegas by checking out a few websites. The
most significant is Poizen Ivy's **www.SinCity
Sounds.com**, an online alt-rock calendar and links resource that's updated weekly. Ivy is clearly biased in
favor of horror-related punk-rock 'n' roll, but her tastes
are broad enough to cover everything from John Doe
to the Hangmen to Steve Poltz. Still, if you're in the
mood for something dark and raw, Ivy's got all the info:
date, time, place and admission price—and some helpful fashion suggestions for upcoming events and reports
on past shows that give you the inside scoop on why

certain ones failed and others succeeded.

Another helpful website for both locals and visitors is **YourLocalScene.com**, a directory of Vegas bands and a concert calendar that has grown to encyclopedic lengths. Here, every band that plays the local clubs with any degree of regularity has a page devoted to it, as well as a link to the band's official website. What's particularly cool about this resource is that you can download mp3s (electronic music files) of the bands before you waste a cover charge on them. The only drawback is that there's way too much chaff and not enough wheat, since any band with $50 can sign up to be included, thereby nixing any editorial selection. There's also a message board that's mostly dominated by shitty local bands sniping at each other's meager successes and failures, but there are also musicians looking to join or form bands. Who knows? If you're a struggling musician, maybe you'll find your new rhythm section in Vegas.

Smash Magazine (www.smashmagazine.com) is a Vegas-based punk-rock/Xtreme-sports magazine with distribution that reaches across the Southwest. This one, too, works as a directory and calendar, but with a strict focus on punk, emo, hardcore and metal. Occasionally, a power-pop group slips in, but this is mostly a testosterone-driven showcase that could use more visible women musicians to make it appealing to both genders. Aggressive-music enthusiasts will totally dig it, though.

For info on upcoming metal shows—death, doom, industrial, glam, gore, goth, grind and power—the best place to look is **www.darksoul7.com**. It's staffed with knowledgeable Vegas metalheads who know their aggro music.

Other, arguably more reliable, resources are the free alt-weekly papers: ***Las Vegas CityLife*** (www.lasvegas citylife.com), ***Las Vegas Mercury*** (www.lasvegas mercury.com) and ***Las Vegas Weekly*** (www.lasvegas weekly.com). The websites are unreliable, so the best thing to do is grab a copy of each of these papers the moment you arrive in town. They're available pretty much everywhere (bus stops, bars, hotel-casinos, major intersections), and they're easier to read once you have them in your hot, little, music-hungry hands.

At least make sure to pick up *CityLife*, since I (like to think I) edit the paper with a particular bent toward whatever's edgy, underground and alternative in Vegas. You know, if I'm still employed there after this book is published.

OK, let's talk venues.

Music Venues

Vegas continues to lose independent, all-ages, non-Strip music venues, typically due to poor management and unfair pressure from Clark County authorities. Despite an unconscious conspiracy to kill the underground music scene, however, there are still plenty of venues in town worth checking out.

The Aristocrat (850 S. Rancho Dr. at Charleston, 702-870-1977) is just a little strip-mall dive bar tucked next to a Smith's. It's a great place to go on almost any night to hear alternative DJs spin garage rock or obscure funk and soul. Indie-rock bands like Eleven Thousand also play here to packed audiences of hipsters, making this a great place to see and be seen in Vegas. Drinks are cheap, too.

Balcony Lights Music & Books (4800 S. Maryland Pkwy. just north of Tropicana, 702-228-2763) is Vegas's only punk-rock bookstore/record shop, and many great indie bands stop to play here on their way to and from L.A.: Joan of Arc, Casket Lottery, Dashboard Confessional, Mirah, Thrice, Saves the Day, Converge. Shows are always cheap ($5), but there are no drinks or food, and no public restroom. (Hey, it's a record shop, not a real venue, remember?) Anyhow, the management here, being pseudo-anarchists, aren't really organized and don't have a website, so make sure and check out an alt-weekly calendar to see who's performing.

The **Cheyenne Saloon** (3103 N. Rancho Dr. at Cheyenne, 702-645-4139) is located way out in North Las Vegas (a town with its own mayor), and it remains a rite of passage for many local punk bands. Interestingly, the place secretly puts on all-ages shows from time to time, corralling the teens in front of the stage and charging them $2 for bottled water. A wretched sound system, but the drinks are cheap and plentiful, which helps dilute the pain.

The **Cooler Lounge** (1905 N. Decatur Blvd. at Lake Mead, 702-646-3009) is a dive bar on the edge of North Las Vegas that's hosted some of the coolest indie and alt-rock bands of the last five years: Wolf Colonel, Jucifer, the Voodoo Organist. There's absolutely no reason to visit the place if there isn't a good music act scheduled to hit the stage, though, so check the listings. Cheap beer is on tap to compensate for the lackluster sound system. Do the dartboards work? No.

In between blasting soccer matches on the TV sets, **Crown & Anchor British Pub** (1350 E. Tropicana Ave.

just east of Maryland, 702-739-8676) offers live music on Saturday nights, and sometimes the bands are pretty damn good, though they're hardly British and rarely sound anything like the Beatles. *Playboy* once deemed this place one of the best bars in the country, and the beers on tap are indeed fantastic. The staff, unlike the bands, is actually British, so make sure you order a Scotch egg and thank them for joining the coalition to liberate Iraq! Just kidding.

The **Double Down Saloon** (4640 Paradise Rd. across from the Hard Rock, 702-791-5775) is where the full-on outlaw rock happens. Any minor-league band whose members are pierced, tattooed and obsessed with horror movies plays this punk-rock Mecca that *GQ* recently deemed one of "The Best Dive Bars in the Country." The jukebox is stocked with tunes by every-one from Sinatra to the Sonics, Black Flag to Bad Brains. Psychedelically inspired murals and a mechani-cal horse add to the crazed ambiance. Video monitors are always playing something weird like *Mighty Joe Young* or Betty Page stripteases. And there's Ass Juice for sale. That's right: Ass Juice. It's got corn in it. Need I say more? Shows here don't begin till midnight, so don't believe any of the posters around town announc-ing 10 p.m. appearances.

Famous John's (252 Convention Center Dr. in the Somerset Plaza, 702-696-9722) is another strip-mall dive bar. The only reason to pay a visit to this place is if a band's playing. Local acts like the Silver State and national indie-rockers like the Album Leaf and Victory at Sea have all put on some tremendous shows here. Check the weekly papers and online resources like SinCitySounds.com to see who—if anyone—might be performing here.

By sheer default, the **House of Blues** (3950 Las Vegas Blvd. S. at Mandalay Bay Dr., inside Mandalay Bay, 702-632-7600) is the best live-music venue in Las Vegas. There have been some absolutely fantastic shows here: Sonic Youth, Modest Mouse, Ben Kweller, David Byrne, Concrete Blonde and on and on. Drinks cost an arm and a leg, so get loaded in the casino beforehand. Stay away from the restaurant, too. Spring for balcony seating, it's worth it.

The **Huntridge Theatre** (1301 E. Charleston Blvd. at Maryland, 702-678-6800) has been under construction, but its renovations are slated for completion soon. It's a big, old, historic building in downtown Vegas that just about every major alt-band has rocked: the Donnas, Interpol, Mogwai. In addition to great international acts, the Huntridge has given many local bands a shot to open for the pros over the years.

The **Hammer House** is where the *real* underground all-ages shows happen. I can't tell you specifically where it is, because the people who run this junkyard—that's right, it's literally a junkyard—on the edge of town will kill me if the cops got word, but if you're looking for some serious hardcore/metal performed by bands of varying quality, then check out www.hammerhouse. cjb.net or scan the flyers at **Big B's CDs & Records** (4761 S. Maryland Parkway just north of Tropicana on the UNLV side of Maryland, 702-732-4433) for more info.

Jillian's of Las Vegas (450 Fremont St. at Las Vegas Blvd. S. inside downtown's Neonopolis, 702-474-4000) has teamed up with *Smash* magazine to become the premiere all-ages punk/ska/hardcore venue in town. In a single month, Jillian's hosted shows by

Tsunami Bomb, Audio Karate, Further Seems Forever, the Kicks, Killradio, the Toasters, Senses Fail, Limbeck, and Q and Not U. If the only band that sounds familiar is the Toasters, then you're probably too old to check out a concert here. But if you have songs by these groups in your iPod, then you're gonna have a blast here. In addition to serving great burgers and such, Jillian's also offers a giant arcade chock-full of video games, skeeball, air hockey and the like. Free parking in Neonopolis's underground parking garage with validation!

The Joint (4455 Paradise Rd. at Harmon, inside the Hard Rock, 702-693-5066) also offers some great shows by established alternative artists like Lucinda Williams and Rufus Wainwright. Yes, it's in the cheesy-ass Hard Rock—bleh. But such is the cultural life in Vegas. Sometimes you have to brave the big, corporate casinos in order to catch a great show, and the Joint has hosted plenty. The sound system here is so first-rate you won't need earplugs—unless you're craving abuse from a metal band, in which case bring earplugs and your mullet. Expensive drinks, so get liquored up at the video-poker machines before entering.

Matteo's Restaurant & Underground Lounge (1305 Arizona St., 702-293-0788, www.matteo dining.com) is in Boulder City and on your way to Hoover Dam, just a half-hour outside of town. The newly renovated lounge is located inside the historic **Boulder Dam Hotel** (built in the '30s), which gives the monthly punk, rockabilly and surf-rock shows there an authentic time warp aura that just can't be faked. SinCitySounds.com mastermind Poizen Ivy promotes many of these shows, so check her website for info.

Moon Doggie's (1750 S. Rainbow Blvd. at Oakey, 702-878-3392) is where the hippies go to listen to "jam band" music on Friday and Saturday nights. Food and drinks are relatively cheap here, though surprisingly there are no vegan or even vegetarian options on the menu.

Pink E's (3695 W. Flamingo Ave. at Arville, 702-252-4666) used to be a pretty cool place to see bands, what with the pink felt pool tables, crazy-looking dartboards and such. Now it's been decorated in a much more reserved, run-of-the-mill sports bar fashion. The music onstage is still a lot of fun, making Pink E's the only spot in Vegas you're likely to find authentic, down-and-dirty blues music.

Roadhouse Casino (2100 N. Boulder Hwy. at Russell, 702-564-1150) is just another cowboy bar (with slot machines, natch) on the far south end of Boulder Highway. In order to draw folks to its remote location, the place has started booking all-ages metal and hardcore shows on weekends. Surprisingly, some pretty good bands have come through here, including acts signed to labels like Metal Blade, Equal Vision and Olympic Recordings. No website, so check the local alt-weeklies or www.lvrocks.com and www.dark soul7.com for info on upcoming shows.

For jazz lovers, there are the smooth-jazz sounds of **Jazzed Café & Vinoteca** (8615 W. Sahara Ave. at Durango, 702-233-2859), which has a bar. But the real reason to hit Jazzed is for its great Italian dishes.

For something a bit more unique in terms of jazz, try **Pogo's Tavern** (2103 N. Decatur Blvd. between Sawyer and Stacey, 702-646-9735), where a group of 80-year-old jazz and big band musicians gathers every

Friday night to jam on old-time standards. Also, the drinks are cheap, and the staff is friendly. Well worth a visit.

Tailspin Bar & Grill (6295 S. Pecos Rd. at Sunset, 702-436-7925) is where the locals go to grab a fistful of hair-metal. A surprising number of headbangers now call Vegas their home, including Vince Neil (Motley Crue) and Kevin Dubrow (Quiet Riot). Not a bad menu despite this joint's (comparatively) bad taste in music. So if you're one of those so-called "hipsters" who wears a Scorpions T-shirt with too much irony, you may actually duck in here and catch a similar-sounding gaggle of metalheads.

Sin City Sounds

Now the tribulation: there are plenty of fun-as-hell, alt-rock bands that will blow your mind with their onstage energy and wicked songwriting.

The scene's most inspiring success story of late comes in the shape of the **Killers** (www.island records.com/thekillers), a retro-'80s, glam-rock quartet that made it big in England before signing to Island Records and barnstorming the States with catchy, pulsing popcraft like "Somebody Told Me" and "Jenny Was a Friend of Mine." Indeed, what made the triumph so sweet—aside from the fact that the Killers were the first Vegas band to make it out of the starting gate since, well, the hair-metal act known as Slaughter—was the fact that the band started out the way most rock bands start: so-so and often shitty. Indeed, the Killers cut their musical teeth playing downtown coffeehouses before graduating to the arenas. Now, of course, the Killers are too big to play an intimate venue in their hometown.

But that doesn't mean there aren't some kickass bands still kicking up dust in the musical landscape. Here are some great alt-rock bands that live—but with any luck will not die—in Vegas, and are likely playing at a dive bar like the Double Down Saloon when you're in town.

Local pop-punk quintet the **Higher** (www.the higher.com) signed with L.A. hardcore indie Fiddler Records, and quickly debuted with a five-song EP called *Star Is Dead*. This is a straight-up emo-pop band that specializes in fierce, twin-guitar riffing and nasal vocals á la Saves the Day. These guys tour their asses off yet always make time for a local show every month.

Gabe Stiff is the Iggy Pop-like wildman who leads the bombing campaigns conducted by his terrific neo-garage band the **Black Jetts** (www.blackjetts.com). If you dig the currently en-vogue sounds of the White Stripes, the Hives—or the eternal majesty of the Stooges—then the Jetts, who are currently signed to Dead Beat Records, will light a fire under your ass. Just be careful about getting too close to the band in a crowded club. Like caged monkeys, the Jetts enjoy flinging spittle, broken drumsticks, microphone stands—whatever's handy, really—at onlookers. This band makes frequent forays into Arizona and California, but always make time for a midnight massacre at the Cooler Lounge on the weekend. Check your local punk-rock record store or look online for either or both of the band's full-length releases: *Bleed Me* and *Right On Sound*.

Need a stoner-rock, Queens-of-the-Stone-Age-type fix? Then look no further than **Bronson** (www.biglizard records.com). Signed to local label Big Lizard Records, this sludge-metal quartet tunes its guitars down to the key of Z and lumbers forward like one of Hannibal's elephants. Sure, the vocals are pretty much just for show, but the monolithic riffs will rattle your skull. Warning: if you forget your earplugs, stuff your aural canals with something—toilet tissue if you have to! **Bronson** is loud. You can probably catch this band opening for WASP at the House of Blues inside Mandalay Bay.

Vegas is blessed—or cursed, depending on the sensitivity of your ears—with one of the most evil bands in the history of metalcore (a super-technical cross between extreme metal and hardcore): **Curl Up and Die** (www.curlupanddie.net). Signed to Revelation Records, CUAD has put forth two stunning collections of "music" in the last two years that are already considered classics within their genre. This band plays a lot of all-ages shows at skateparks, so if you wanna see how, in fact, the kids are all wrong with their insane moshing, then let CUAD restore your faith in the notion that, when left unsupervised, kids begin to mimic the characters in William Golding's *Lord of the Flies*.

Although nowhere near as visionary as CUAD, there's a more traditional hardcore act called **bydeathsdesign** (www.bydeathsdesign.net) that offers the usual: screaming, metallic riffing, and pummeling drums. Signed to local Embryo Records, bydeathsdesign is lyrically ambitious despite their inarticulate-sounding rage. So if you like Greek myths, theories of evolution and philosophical meanderings laced with blood-and-guts guitarmanship, then go see this band tear it up at Jillian's inside Neonopolis.

For those who find themselves drawn to Foo Fighter-ish rock, the band **Magna-Fi** (www.magna-fi.com) delivers crushing guitar-pop in spades. Signed to Phoenix indie Aezra Records, this quartet has a searing collection of radio-ready tunes called *Burn Out the Stars* that delivers pure, hook-laden rock. Magna-Fi secured a spot on last year's Ozzfest, establishing a fanbase outside of Vegas and making a name on the international scene as well. (Interestingly, Magna-Fi sells more CDs in Japan than in the States.) You can catch these guys tearing the roof off Pink E's and the Lounge inside the Palms on weekends.

There's a softer side to the Vegas music scene that's evident when you listen to the **Bleachers**, a local alt-folk duo consisting of singer/songwriter Joe Maloney and guitarist/producer Marco Brizuela. Firmly entrenched in the stylings of artists like Nick Drake, Elliott Smith and Palace, the Bleachers are the flagship band of **Village Industries** (www.village-industries.com), a label that sprung up on the scene in 2003. The only other artist signed to the fledgling company is the mysterious songstress **Orange Sheila**, whose elliptical songs bring to mind Cat Power. You can see the Village Industries artists perform at coffeehouses like the Iowa Café (300 E Charleston Blvd, 702-366-1882) and the Pride Factory.

The **Nines** are the best surf-punk-lounge-jazz instrumental combo you'll find anywhere—period. Comprised of members of various other bands, the Nines never cease to amaze listeners with their ability to effortlessly vamp from a Cole Porter standard to Pink Floyd's "Comfortably Numb" to the Clash's "London Calling." If God (or Satan) needed a band for a Vegas shindig, these are the guys He'd hire. Infinitely cool, immaculately dressed, the Nines bring aesthetic sophistication and a retro-sensibility to every song they cover. Difficult to catch due to the members' commitments to other, more commercial musical endeavors, this is the band you should hunt down in the music listings of the alt-weekly papers in Vegas.

Got goth? If not, you might want to take a sip of **Moonvine**'s dark brew. Think Sisters of Mercy with lots of piano and violin. Be sure to see them blocking out the desert sun with their minor-chord melancholy at events like the annual Avant Garden event here in town. To find out where Moonvine—or any other Vegas goth band is playing for that matter—check out www.vegasgoths.com.

Black Camaro is the most consistently stoned band in Vegas. Sixties psychedelic rock meets Ween with this savagely demented rock quartet that borrows and steals props to make their live shows akin to a fireworks display. (Don't worry: the band is no Great White!)

The **Loud Pipes** (www.theloudpipes.com) are the greasiest biker-metal band on the planet, and if it weren't for the poutingly pretty Roxie on bass, they'd be the ugliest. Think Motörhead crossed with the MC5 with some shrieking vocals, and that's what the Loud Pipes deliver. Despite its awesome prowess, this group isn't afraid to play a house concert or a middle-of-the-desert show for music-hungry teenagers. Once this band figures out how to present itself in a slightly more professional manner (that is, when they stop hocking loogies at the audience), the major labels will come a'knockin'. Check the Loud Pipes' website for their next show. You won't be disappointed. The band is energetic, pure evil and ready to erupt. Bring earplugs.

The **Pervz** (www.thepervz.com) are an old-school,

punk-rock power trio that delivers the hooks in skinny-tie spades. Two brothers from Texas and a Chicago-born-and-bred drummer pack the Double Down night after night with catchy rave-ups like "Pieces of You" and "I Don't Wanna Hear It," all of which bristle with undeniable energy and attitude, putting most other bands—signed or otherwise—to shame. Three chords, the ugly truth about boys and girls, and a hook—it's all any live music fan needs. Bring earplugs.

Killian's Angels (www.killiansangels.com) are the best all-female Irish/Celtic band in the country. This band, consisting of acoustic guitars, a violin and a tuba player, does everything from Peggy Lee's "Fever" to the Eurhythmics to original compositions that will make you wanna drink harder, faster, pussycat, kill kill! Killian's usually signs up for month-long gigs at different casino bars across the valley, so check their website to find out where and when they're playing before you get to town. This is a seriously fun band.

Some other Vegas bands worth checking out: Grindcore trio **Guttural Secrete**, whose unique brand of noise is reminiscent of Pig Destroyer and Circle of Dead Children; **Amber Halo**, an alt-rock trio whose recent Big Lizard debut, *Stealing Insulin*, has gotten rave reviews from the local media; **FFI**, another cornerstone of the Big Lizard roster whose punk-meets-classic rock songcraft has endeared the band to many critics and fans; the **Vermin** (www.myspace.com/thevermin), an old-school punk-rock trio with three-chord songs like "She's So Obscene" and "Shit Talkin'." (Warning: The Vermin spit beer at the audience. Bring an umbrella.) **Eleven Thousand** (www.eleven thousand.com) specialize in an abrasive, guitar-heavy brand of indie-rock that might melt your ears off if you're not too careful. **The Silver State** (www.silver statetheband.com) is another indie-rock trio that can stand toe-to-toe with any of today's best bands. **Psychic Radio** (www.embryorecords.com/psychic radio.htm), meanwhile, blends Beatlesque hooks with punchy guitars and pummeling drums. **The Utmost** (www.utmostmusic.com) is utterly appealing, offering rowdy pop-punk songs that'll make you hum along even as you're throwing those devil horns in the air. **Big Friendly Corporation** (you can download songs at Mperia.com) is a cross between Sparklehorse and Modest Mouse, so don your best vintage/thrift-store threads before heading out to see this remarkable quintet. **Judee's Pryde** adds a little GNR to the alt-rock mix; check this band out if you're able—they're accessible yet edgy.

Vegas Vinyl

For underground and alternative hip-hop and rap vinyl, check out the **HipHopSite** (4700 S. Maryland Pkwy. just north of Tropicana across from the In-N-Out Burger, 702-933-2123, www.hiphopsite.com). From MF Doom to RJD2, from Terror Squad to Young Buck, from Masta Ace to Shyne, this little shop next to the UNLV campus carries it all, including new and used tapes and CDs. They usually advertise with a discount coupon in some of the local papers like *Smash*.

Balcony Lights Music & Books (4800 S. Maryland Pkwy. just north of Tropicana, 702-228-2763) has a limited but thorough selection of punk-rock/hardcore/metal/reggae vinyl, including full-lengths, 7 inches and 45s. Nothing here is close to "collector quality," but if you absolutely need some Bad Brains, Circle Jerks or Fear in your musical diet, this is the place to hit.

Big B's CDs & Records (4761 S. Maryland Pkwy. just north of Tropicana, on the edge of the UNLV campus, 702-732-4433) has it all, baby: Alice Cooper's *Constrictor*. Dave Pike's *Limbo Carnival*. Little Willie John's *Fever*. And—for the kids—Sublime's *40 oz. to Freedom* picture disc. If you're looking to spin the black circle, you'll need to stock up at Big B's, where vinyl is still king (and Morrissey is still queen). At first, the employees may seem a little snobby, but that's only because they've got a wealth of alt-rock, indie-rock, hip-hop, avant-pop information packed into their craniums that goes to waste whenever someone asks for the new Toby Keith or DMX. So go ahead and ask 'em about that weird old song or band you heard on the Internet or when you were driving to San Diego to see your ex. Chances are they've got the vinyl.

Wax Trax Records, Inc. (2909 N. Decatur Blvd. between Desert Inn and Sahara, 702-362-4300) is where serious collectors go to find pristine, scratch-free vinyl. You'll also find plenty of rare items (45s, 78s, cassettes, CDs, 8-tracks) and memorabilia, including the Beatles' controversial butcher cover for *Yesterday and Today*, as well as Redd Foxx's X-rated records. If you're a "junker" (someone who sifts through piles of junk for that elusive vinyl diamond in the rough), don't even bother walking in the door. Wax Trax doesn't offer anything for less than $20. So draw up a list before you head on over.

Musicians Wanted

Hard rock and nü-metal dominate the radio airwaves and concert halls, and the more proficient performers have already sold their souls and firstborn children to the Strip casinos, where they're doomed to play cover songs until they die. So if you move to Vegas and need to play to keep your sanity intact, look in the Musicians Wanted ads in the back of the free weeklies, or check out the local websites with message boards contianing calls for musicians that may help you find that perfect drummer or harp blower: www.yourmusic scene.com, www.darksoul7.com, www.smashmagazine .com. You can always walk into any of the rehearsal studios, record shops and music stores listed below and scan the bulletin boards.

Practice Makes Perfect

Robert Allen Studios (3977 Vegas Valley Dr. at Lamb, 702-431-8441) is located behind Boulder Station off Boulder Highway. Ex-Guns N' Roses drummer Steven Adler practices there with his band, along with many other washed-out glam rock icons from the '80s who now play the redneck-metal circuit up and down the casinos on Boulder. But many Vegas indie-rock acts hone their chops there, too. And if you can find a band with which you share a certain sound, they'll probably let you split the rental cost of the space.

The Alamo Rehearsal Studios (310 W. Utah Ave. at Oakey, 702-382-8707) are located downtown. This is where Vegas's loudest metal/hardcore/noise outfits learn to split eardrums. Essentially an old motel, the Alamo will cause shivers to go up and down your spine as you walk through the halls, a new and horrifying sound erupting from behind each door. Yeah, it sounds like a haunted house, but the guy who works the front office is a no Crypt Keeper. Don't let his tats and massive frame (or that book about how to become a Navy Seal) intimidate you.

There are rumors of "good but expensive" circulating about **On Stage Rehearsal Studios** (3905 W. Diablo Dr., #103 between Hacienda and Russell off of Valley View, 702-798-4141, www.onstagerehearsal.com), located just seconds from the Strip. In order to keep you pampered like a poodle-haired glam-rock band from the '80s, this studio features PA systems, lighting, air-conditioning and sound-proof walls. There are even mirrored walls for worshiping your badass, rock 'n' roll self. Is it all worth $16 an hour (three-hour minimum)? Um, if you're Poison, then maybe.

Instrumental Success

There are several great places to buy instruments. The first is a vintage shop called **Cowtown Guitars** (2797 S. Maryland Parkway at Karen, 702-866-2600), where you can buy just about any kind of classic guitar, from Sunburst Les Pauls to the original Gibson SGs. There are any number of guitar and bass amplifiers, from Suns to Marshall stacks to Ampegs, and a nice selection of vintage effects pedals (Superfuzz? Big Muff?). And if you need your vintage gear fixed, Cowtown has the best techs in town. Music lessons and consignments are also available.

No drums or keyboards, though, so you may have to go next door to the **Sam Ash Music Store** (2747 S. Maryland Parkway at Karen, 702-734-0007) for all your non-guitar needs. Sam Ash bought out Mars Music, which went bankrupt a few years back, so you get the idea: a vast display of brand-spankin'-new axes, amps, kits and keyboards. Also, there's a cornucopia of music instruction books, tapes and DVDs, as well as transcriptions of every must-have album, from *Sabbath Bloody Sabbath* to *Nevermind*.

Guitar Center (3085 E. Tropicana Ave. at Pecos, 702-450-2260) is another superstore that's really no different than Sam Ash, except a lot of local musicians work there, and if you're indie-rock enough in their eyes, they'll cut you a decent deal. This place has a wide selection of digital effects.

If an ax is all that's on your mind, **Ed Roman Guitars** (4305 S. Industrial Rd. #165, just north of Tropicana, 702-798-4995, www.edromanguitars.com) has more than 3,500 guitars in stock. Just three minutes from the Strip and located right next to a rock-themed bar and grill called TommyRocker's, Ed Roman also does trade-ins, free appraisals and on-site repair. Their website is huge, so if you like to windowshop for something like a BC Rich Warlock bass, this is your kind of place.

There's also **Advanced Guitar** (3340 E. Tropicana Ave. just west of Pecos, 702-450-6161), which primarily sells rock 'n' roll axes. And there isn't a better place to shop for acoustic six-strings—Heritage, G&L, Silvertone, Samick, Tribute—than **Vegas Guitars** (3051 Coleman St., 702-251-1580, www.vegasguitars.com). The website for this place is immense.

If guitars and drums just ain't your bag, try the stores that specialize in traditional high school marching band style instruments like **Family Music Centers** (Westside: 8125 W. Sahara Ave. at Cimarron, 702-360-4080; Eastside: 2714 N. Green Valley Pkwy. at Sunset, 702-435-4080), where you can find acoustic and digital pianos, band and orchestra instruments and a mess of sheet music. There are also a few **Kessler & Sons Music Stores** (check www.kesslermusic.com for the nearest location), as well as a **Music World Inc.** (2295 E. Sahara Ave. at Eastern, 702-457-6869) specializing in keyboards.

And don't forget to check the message boards at Vegas music sites for instrument sales: sometimes bands unload their gear for a song.

Recording Studios

Need to commit those nasty little rock songs you penned in your seedy motel room on Fremont St. to tape or, more accurately, to some kind of pro-quality digital recording? Well, **Alien Audio** (4405 E. Sahara Ave. between Sunset and Eastern on the east side of the airport, 702-641-7716) is good enough for Insane Clown Posse, who recorded a few tracks from their last album at this studio that borders the airport.

Platinum Sound Lab (6000 S. Eastern Ave. at Patrick, 702-795-1750, www.spin78.com) is where pop bands like Chicago have recorded, but don't let such slick clients fool you. The engineer there spent some time working for the now-defunct Westwood Studios videogame designer company. And that was before he worked with Whitney Houston, Kenny G and Lionel Richie. Expensive, sure, but it's easy to see a Vegas alt-rock band making a demo here that lands them a major-label deal.

The **Tone Factory Recording Studios** (2300 Patrick Lane at Eastern, 702-795-3886) is where pop-rock maestro **Brian Jay Cline** (www.brianjay cline.com) — who rarely performs live these days — creates his catchy Marshall Crenshaw-meets-Dave

Alvin albums. The head engineer there is a musician—
a drummer—so he knows how to capture great sounds.
Probably the most reasonably priced place to cut a
record.

CD Mastering and Manufacturing

Formerly Tom Parham Audio Productions, now
known as **Odds On Recording** (14 Sunset Way at Sunset Rd., 1-877-633-7661, www.oddsonrecording.com)
in Henderson, this is where many metal bands have
mass-produced their demos over the years. Parham also
runs the local label Embryo Records.

Digital Insight Recording Studio (2810 S. Maryland Pkwy. #C between Sahara and Desert Inn,
702-792-3302, www.digitalinsightrecording.com) has
pretty good rates for CD duplication that many local
groups and national artists (Missy Elliott, Faith Hill,
Celine Dion, et al.) have taken advantage of. This studio also has a "local band rate" for studio time, though
I wasn't able to find a single local band that had recorded anything at Digital Insight. You can also get your
CD mastered here.

One Vibe Studios (3696 E. Russell Rd. between
Pecos and Sandhill, 702-436-3055, www.onevibe
studios.com) supposedly also has pretty good duplication and mastering prices. As far as recording goes,
again, I had trouble finding any local alt-rock/indie
musicians who had used the place.

Open-Mic Nights

Open-mic nights in Vegas can get especially grue-
some, given that Sin City is becoming a refuge of sorts
for washed-up L.A. glam-rock bands from the '80s.

Your best bet is to check online and flip through the weekly papers to get an idea for the open mic that best suits you. There is a pretty decent Chicago-style blues jam every Monday night at 10 p.m. inside the **Double Down Saloon** (4640 Paradise Rd. across from the Hard Rock, 702-791-5775, www.doubledownsaloon.com), where you can wank your six-string ax to your heart's content.

Booking/Promoting a show

Are you in a band that wants to play Vegas? Well, most of the venues listed above are fine to contact directly, especially Balcony Lights Music & Books, the Cooler Lounge, and the Double Down Saloon. Just send them a coherent press kit with a CD and you should be set. Don't expect any help with promotion, though. You'll have to handle that on your own by contacting the free weeklies. And they're completely buried under an avalanche of out-of-town bands' press kits.

So anything fancier than that requires a booking agent (who'll take a cut of ticket sales), and Vegas has only a few who put on shows regularly. First, there's **AdvanceWest** (www.advancewest.com), a company that's brought in some great alternative and indie-rock acts over the years, from Apples in Stereo to New Bomb Turks. Contact them via the website.

Then there's **Revenge Therapy Productions** (www.revengetherapy.net). If you're in a hardcore, metal or noise act—or anything in between—this is who you need to work with to ensure your show actually happens and gets promoted. Also, these guys have close ties to the Loud Pipes, a popular underground bikercore band that will always draw a crowd.

If you already have a show booked and all you need is promotion, you should contact the good folks at **Shoestring Promotions** (www.shoestringpromo tions.com, 702-360-7881). These people will plaster the cars in every parking lot in town with fliers announcing your band's imminent arrival. And they'll hit every record store and coffeeshop, too. So if you want the music-hungry kids of Las Vegas to know about your show at Jillian's, get Shoestring involved.

If these three booking/promotion outfits won't get on board, then you can always beg Ryan Kinder at **Big Lizard Records** (www.biglizardrecords.com) to book and promote a show. He's always swearing off this kind of work, but if your band is good enough (like, if your band happens to be Camper Van Beethoven), he'll make an exception and suddenly emerge out of retirement. Just don't suck, OK?

And Poizen Ivy over at **SinCitySounds.com** is always perfect in a pinch. Again, it really helps if you don't suck.

Remember: booking/promotion costs money! God knows these folks ain't gonna rip you off—after all, they do what they do 'cause they love music and not because they wanna get rich. So when they ask for a little money up front, don't freak out, please. They probably just need to cover the Kinko's tab.

Tattoos, Etc.

Hell, you're gonna need some loud tats and piercings to go along with all the loud music. So why not head over to Dirk Vermin's **Pussykat Tattoo Parlor** (4972 S. Maryland Pkwy. at Tropicana, 702-597-1549). Vermin, you should know, is a punk-rock institution in and of himself. His eponymous band, the Vermin, have been pounding out three-chord anthems to beer and nihilism since 1984. And he's been a tattoo artist for just about that long. If you're looking for an old-school tat (mermaid, Mom, Mephistopholes) or a retro-pop job (Batman, Bettie Page, Betty Boop), this is where you need to go. And there's an impressive selection of biker-style demons, skulls and monsters as well. Piercings are also done here: inner labial, anyone?

Plenty of local musicians swear by **Diversity** (multiple locations). One man's tattoo is another man's mistake. So before you pay good money to have a screaming skull permanently inked on your ass, make sure you have a clear head. But if you're going to spend your money on a tattoo, Diversity is the place to go. With five locations in Vegas, this chain has the largest selection of tattoos and body jewelry in town—more than 1,000 selections. The store also prides itself on its body-piercing studio.

Tattoo styles include black and gray, Celtic, tribal, custom work, and new school. While you're at it, what better souvenir to remember your stay in Vegas with than with a metal hoop through your scrotum?

Sin City Tattoos (102 E. Charleston Blvd. at Maryland, 702-387-6969) is the downtown shop across from the Huntridge Theatre. It looks scary and hardcore from the outside, but once you're inside, the folks are friendly and competent.

NIGHTCLUBBING

If you wanna hit the Strip and dance the night away to mainstream music or the endless techno-deep-jungle-trip-hop beats of some overpaid DJ who is sharing his electronica/dance music collection with hordes of mindless, meat-shopping hordes, then by all means, go for it. Pick up any alt-weekly and you'll find plenty of sexy ads designed to lure you in. But after spending $300 with no action to be had, please understand that Vegas nightclubbing, however spectacular it may seem, is always the same: expensively heartbreaking.

The best place to casually enjoy some full-on funky jams without having to empty your entire wallet is **The Get Back**, a monthly dance party at the **Icehouse Lounge** (650 S. Main St. at Bonneville, 702-315-2570). DJ John Doe spins the rawest funk and soul from the '50s, '60s and '70s. The event takes place at 10 p.m. the night of each First Friday event. Look sharp (but not cheesy sharp), and—brace yourself—there's a $5 cover.

Open Forum at **Studio 54** (inside the MGM at Las Vegas Blvd. S. and Tropicana, 702-891-7254) happens once a month, usually on the third Saturday of the month. Check the weekly papers and the club's website for more info on this party that involves Blue Man Group musicians and pro dancers. It's like a mainstream nightclub turned free-for-all circus.

But if you just wanna wear blue jeans and forego the whole nightclub wardrobe, grab your dirty Chuck Taylors and head to the Double Down any Monday night at 10 p.m. for **Branded**, orchestrated by the **Bargain**

DJ Collective, a group of punk-rock fans turned vinyl enthusiasts who'll spin anything your retro-heart desires. These guys don't have a website, but you can keep tabs on them by checking out Ryan Kinder's label site, www.biglizardrecords.com (Kinder spins indie and alt-rock classics under the rubric "DJ Tiger"), or www.doubledownsaloon.com.

Punk Rock Burlesque

The bump'n'grind of our grandfathers' generation is back, and what better place than Vegas to enjoy such a raunchy renaissance?

The two best strip troupes are of the punk-rock/goth stripe: First, the **Sin City Grind Kittens** (www.sincitygrindkittens.com), who are fleshy, flashy, and full of fun. You can catch the Kittens performing at events like psychobilly festivals and hot-rod conventions. Second, the **Babes in Sin** (www.babesinsin.net) will wow you with a jaw-dropping performance at places like the Cooler Lounge or downtown in the Arts Factory on a First Friday. If the troupes are dancing on the same night, pick one and you can't lose. After all, pasties and punk rock go together like peanut butter and chocolate, right?

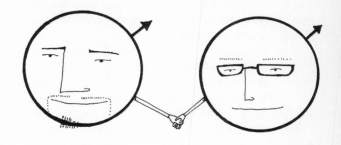

QUEER VEGAS

Las Vegas could be characterized as the adolescent daughter of the U.S.A. since it exhibits the full spectrum of good and wicked behavior depending on the day, hour, or minute. A little bit naïve (believe it or not) yet brazen, and appearing more worldly than she really is. With more than enough faux bravado to spare, full of energy, grand plans and clever schemes, though precious few may become reality. Ambitious but a little lazy... too much fun, exciting to be with, and extremely annoying at times.

Vegas and its Gay/Lesbian/Bisexual/Transgender (GLBT) community are extreme like this in every way.

Thriving on the dramatics of an ultra-fascinating existence, totally self-indulgent, quintessentially narcissistic … and then unexpectedly poised, even gracious for a moment, even with that oh-so-awkward gait. Not yet grown into a mature exterior, but yet so stunningly beautiful in a sparkly new dress.

Las Vegas is the Land of the Drag Queens and the New Entertainment Capital of the World! Our GLBT icons of glamour and grace perform nightly in every show on the Strip. Gay entertainment denizens Siegfried and Roy have called Vegas their home for decades. Though never openly gay to the public, everyone knew; and their highly visible, 40-plus-year partnership was so hugely successful that they're regarded by many as the entertainers who made Vegas the entertainment Mecca it is today.

Lest we forget Vegas's favorite son, Liberace, you truly must make the only gay pilgrimage to that homage to excessive fabulousness, the **Liberace Museum** (1775 Tropicana Ave. at Spencer, 702-798-5595, www.liberace.org). While you're there, pop into **Goodtimes** for a drink (same shopping center), with the best gaming and promotions at a gay bar in Vegas. They also have a kickass sound system, state-of-the-art lighting and the only stainless-steel dance floor I've ever heard of.

And now gracing our stages in the wake of Roy Horn's tragic career-ending accident, we have **Sir Elton John** and **Celine Dion**; and that delicious Diva of Wigstock fame, **Joey Arias**, who hosts Cirque du Soleil's new controversial show "**Zumanity**" at New York-New York hotel-casino.

In fact, we've been headlining Vegas for decades.

One more historic note: Breck Walls' **"Bottoms Up"** is one of the longest running shows in Vegas at the Riviera hotel-casino. You can also catch **"La Cage"** starring the infamous (she sued him) Joan Rivers impersonator, Frank Marino, and his cast of superstar female illusionists there. Incidentally, the Riviera was the first hotel-casino in Vegas to host and advertise openly gay events in a series of gay comedy shows called **"Gaylapalooza."** The Las Vegas GLBT Pride Association worked in conjunction with then-Riviera Entertainment Director Steve Schirripa (recently seen playing Bobby on "The Sopranos") to produce these landmark events for gay Vegas in the late 1990s.

So you want to come see how Queer Vegas plays in its own backyard? We're not so different than the rest of the homosexual world, really. Except for the fact that you can gamble at almost any location or activity you attend, including the Laundromat and every 7-Eleven in town!

The specter of the gambling industry here, whether you ever choose to partake of it or not, affords individuals a kind of handy conversational element. It can inspire instant camaraderie, as well as suss out the financially (if not spiritually and socially) tenuous potential tricks whilst you cruise Sin City in search of some sinning of your own.

As with all settlements of humans on the planet, there exists the typical buffet, if you will, of gay bars. Now please be fair: do not compare Vegas's Gay scene to any major city such as San Francisco, Chicago or New York. Come and enjoy what we have to offer as a city that has a growing GLBT community, and each time you return you'll be amazed at the great changes that have taken place.

There are currently about a dozen gay bars, most of them friendly to both men and women; all of them provide the typical array of entertainment and entice-ment to go. Count on finding a veritable plethora of beer busts and go-go boyz, karaoke and amateur strip contests, fundraisers and drink specials, ladies nights and underwear nights (for men), dancing, cruising, and yes, gambling, gambling, gambling.

Special events, fun things to do, and the locations of all gay-owned and gay-friendly businesses are al-ways listed in detail in the *Las Vegas Bugle* (now rechristened *Bugle Q Vegas* and online at www.lasvegasbugle.com). Sin City's only gay monthly publication, *The Bugle* is your best resource guide to what and who is currently happening here socially, po-litically, and spiritually. Pick up a free copy at the bookstore, **Get Booked** (4640 Paradise Rd. just south of the Hard Rock, 702-737-7780), which is located in what is affectionately called the Fruit Loop.

The Fruit Loop contains the largest and most var-ied concentration of gay bars in the city. Next to Get Booked, you'll find the **Buffalo** (702-733-835), Vegas's legendary leather bar, where women and men are just as likely to leave with a wonderful recipe for cheese-cake as a neat row of welts on their asses. The **Freezone** (610 E. Naples Rd., two blocks south of the Hard Rock, 702-794-2300) is just across the street and features a ladies night on Tuesdays, a respectable dance floor, res-taurant, and a mixed crowd the rest of the week.

On the opposite side of Paradise Road stands the **Gypsy** (4605 Paradise Rd. just south of the Hard Rock, 702-731-1919), remaining Vegas's only true gay dance club, disco-style for more than 20 years. Don't worry, she's had a face-lift or twelve in the meantime, Girrrl!

Scattered about town are several establishments that are primarily male-oriented in their activities and clientele. Not that ladies would be made to feel unwelcome. However, something along the lines of an "Underwear Night" isn't typically on a visiting lesbian's must-see list in Vegas. But if an underwear night and the like are what you're looking for, be sure to check out the **Las Vegas Eagle** (3430 E. Tropicana Ave. at Pecos, 702-458-8662), the **Spotlight** (957 E. Sahara Ave. at Maryland, 702-696-0202), **Snicks** (1402 S. 3rd St. at Imperial, 702-385-9298), the **Backdoor** (1415 E. Charleston Blvd. at 15th, 702-385-2018), **Badlands** (953 E. Sahara Ave. inside Commercial Center just west of Maryland, 702-792-9262), and Vegas's first exclusively gay resort, the **Blue Moon** (2651 Westwood Dr., reservations and information: 800-851-1703). All of these places also feature the aforementioned beer busts, fundraisers, karaoke, etc., too.

The **Las Vegas Lounge** (900 E. Karen Ave. Ste. H, 702-735-0885), Vegas's only tranny bar — that admits it, anyway— is friendly and open to all. It's located in Commercial Center and features (what else?) the most fabulous drag shows in Vegas every night at 10 p.m.

Other mixed clubs in town with dancing, karaoke, and the rest of the buffet include **Flex** (4347 W. Charleston Blvd. inside Commercial Center just west of Maryland, 702-385-3539); and **Goodtimes** (1775 E. Tropicana Ave. at Spencer, 702-736-9494). **Backstreet Saloon & Dancehall** (5012 S. Arville Rd. at Tropicana, 702-876-1844) is home of the Nevada Gay Rodeo Association (NGRA) and hosts line dance lessons Thursday nights for men and women (always well-attended by Vegas cowgirls and their admirers), and NGRA parties and fundraising events monthly.

Las Vegas's latest gay club—and the first that fronts the Strip—has taken up residence at the Desert Passage inside the Aladdin: **Krave** (3663 Las Vegas Blvd. S. at Harmon, 702-836-0830), a large dance floor with surrounding booths, multiple bars and a separate room suitable for a restaurant. Helmed by Sia Amiri (former owner of the popular Rage in West Hollywood) and "the high priest of gay parties" Jeffrey Sanker, Krave offers a sensual environment with sheer curtains, cozy nooks and plenty of hot go-go boys to keep the crowd entertained. Saturdays are "Everything You Desire" nights, or, as the club puts it, "an evening of excitement and debauchery." I can identify with the debauchery part. Those looking for a little more hedonism should come out for the *Fashionistas*, a show that takes its inspiration from adult films, Mon.-Sat. at 8 p.m. (Tickets prices start at $49.) The club also includes its own restaurant, E.A.T. Krave is open Fri.-Sat., 10 p.m.-6 a.m and Sun., 10 p.m.-4 a.m. $10 for locals; $20 out-of-towners.

A special note to all the lesbians and bi women: Please don't be discouraged by the apparent lack of (specifically) women's venues. Most of the locations described as "mixed" have an attractive crowd of women of all ages, both locals and visitors, especially on weekends and special events. Get more information specifically for women visiting Vegas at **www.bettys outrageousadventures.com**.

Located in Commercial Center, which is fast becoming something of a second Fruit Loop, is a new coffeehouse and pride memorabilia store chain called the **Pride Factory** (953 E. Sahara Ave. Ste 1B, 702-444-1291). This is turning out to be a great meeting place for gays, lesbians, trans, bi, and our friends of all ages. The Pride Factory is open 24/7 and is located in the same shopping center as the **Las Vegas GLBT Community Center** (953 E. Sahara Ave. Ste B-25, 702-733-9800). In existence for more than 10 years now, the Center hosts meeting of all GLBT community organizations and clubs, offers a vast array of monthly and weekly meetings, classes and workshops, AA and NA meetings daily, youth outreach, HIV/AIDS services and is simply a great resource for anyone visiting or moving to the greater Las Vegas area.

For those of you who need a good dose of spirituality in the face of so much unbridled sin, the **Las Vegas Metropolitan Community Church** (1140 Almond Tree Lane and Maryland, 702-369-4380) meets every Sunday at 8 a.m., 10 a.m., and 11:30 a.m. and Wednesdays at 6 p.m.

Gay marriage? Sorry, not recognized by the state of Nevada. But the **Viva Las Vegas Wedding Chapel** (1205 Las Vegas Blvd. S. between Oakey and Charleston, 702-384-0771, www.vivalasvegas weddings.com) has a non-discrimination policy that allows same-sex couples to receive the same hospitality and commitment ceremonies that more traditional couples are allowed.

As Vegas has begun to grow up and our population becomes more sophisticated (both in its behavior and its belief systems), we're seeing the growing influence and visibility of gay culture emerging here as well. Clubs, events and local entertainment venues are certainly more noteworthy and are drawing a more diverse, desirable mix of people. Just remember: Vegas is still the City of Sin, where only *certain* sins are acceptable to all, and we still have a lot of growing up to do. So be out loud and proud, but at the same time be prudent about where you do so—and with whom.

LITERARY LAS VEGAS

While you're destroying your body and soul in Vegas, you might as well corrupt your mind with some books that do much to further this town's reputation as a sinister bacchanalia. Here are a few choice tomes that will help summon the dark energies necessary before saddling up to that slot machine/bar/tranny stripper:

Though *Everybody Smokes in Hell* (Ballantine Books, $14) isn't widely considered to be a novel about Vegas, guess what? It is. Ping-ponging between Vegas and its equally depraved sister city, Los Angeles, author John Ridley serves up a cast of characters rendered with such fierce authenticity that the reader can't help but laugh as he feeds them into the meat grinder. Drug dealers, hit-women, lawyers, and a 7-Eleven clerk all struggle to get what they got coming to them in this hardcore tale of murder, revenge, and pop music. Pick up a copy and make room in your schedule; you won't stop reading until the last page has been turned. Also contains the funniest suicide scene ever in literary history.

For natives, it's hard to imagine a more fully realized rendition of the city's past than in H. Lee Barnes' *The Lucky* (University of Nevada Press, $22). Although the book is largely focused on the coming-of-age story of its protagonist, the star of the novel is Vegas itself. Never before has the City of Sin been rendered more authentically, down to the (now long-gone) A&W Hamburger stand on East Charleston and sunbathing chorus girls; it's all there. Little details like these make the book stand out from the vast morass of bullshit fictionalizations that consistently fail to get the facts straight. Highly recommended.

One of my all-time favorite books about Vegas is (of course) ***Fear and Loathing in Las Vegas*** (Vintage Books, $12). Hunter S. Thompson's 1971 classic pastiche of drug-induced psychoses served up as tripped-out travelogue provides the basis for one of the most interesting fictionalized accounts of the town from the perspective of a tourist. The surreal nature of the city combines with the drug elements into one of the most difficult to put down, piss-your-pants-funny accounts of the town I've ever read. If you can't get enough of the film, by the way, check out *Where the Buffalo Roam* (1980) starring Bill Murray and Peter Boyle; it's excellent as well.

Novelist Heather Skyler's ***The Perfect Age*** (Norton, $24.95) offers a classic premise: a young girl, Helen, working as a pool lifeguard at the Dunes, comes of age in the desert heat of Las Vegas. Her relationship with her sweetheart Leo unravels when she meets an older boy named Miles, whose confidence and good looks draw her to him. At the same time, her mother, Kathy, struggles with the monotony of her marriage to an uptight UNLV professor and begins an affair with the looser, more freewheelin' Gerard, Helen's boss. In the course of three summers, mother and daughter must navigate life's disappointments and love's petty betrayals. Ultimately, they discover that no matter one's age, there are always lessons to be learned and mysteries that will never be solved.

And, finally, another literary offering that was made into a major motion picture: John O'Brien's unflinching journey into boozed-out depravity and death, *Leaving Las Vegas* (Grove Press, $11). O'Brien (who took his own life shortly after the sale of the movie rights) will crush your soul in this tale of one man's slide down the razor blade of life. Vegas serves as the perfect setting for the book's endless series of empty bottles, emotional bruises, and desperate, tragic nights spent with a bluesy hooker, heart of gold included at no extra charge.

Poetry Readings, Slams, Etc.

Stop by and ask about days, times, etc. or visit www.localendar.com/public/VegasPoetry for details.

reJAVAnate (3300 E. Flamingo at Pecos)
Coffee Bean & Tea Leaf (4550 S Maryland Pkwy)
Sweet Georgia Brown's (2600 E Flamingo at Eastern)
Norma Jeane's (3650 S. Decatur at Twain)

COOL DIVE BARS IN THE INFERNO
and a few notes on gambling

While Vegas offers a plethora of liquor-based pos-
sibilities, let's assume that you're more than capable of
finding the bar located within your hotel: all the major
casinos maintain multiple drinking establishments
within easy reach of the gaming floor (in the hopes that
their patrons will be as drunk as possible while throwing their money away). And if you're into wasting your
money stalking celebrities, then visit **Ghostbar** (at the
Palms, 4321 W. Flamingo Rd, 702-938-2666). But if
you're looking to do some serious drinking, let's hit
the town.

Before we begin, however, some general advice
(and a few key tips) are in order.

How to Drink for Free (in Bars)

The reason that there are so many bars in Vegas is pretty simple: the owner only has to have five video-poker machines installed to cover the cost of rent for each night. And that's figuring the night will be slow. Consider that most places in town have more than 15 machines operating on the bar, and that most people who come in will lose somewhere between $20 to $100 during any given night, and the relatively hidden economy of scale behind the place begins to become apparent. Factor in one or two serious gamblers and you're in business.

So what does this mean? It means comps: drinks on the house, complimentary, in exchange for throwing your money away on video poker. While a bar's policy on comps varies from place to place, here's a general outline of what to expect:

Understand that, essentially, there are only three kinds of bars in Vegas: gaming bars (places that are designed to make all their money from clientele's gambling losses), dive bars (places of character that make their money any way they can), and specialty bars (places that cater to customers looking for atmosphere more than gambling or even drinking). The Westside of town has most of the best and easy-to-find gaming bars, although they can be found all over town (Green Valley is a close second). The central and university-district areas have the best dives. Of the three categories, your best bet for drinking for free are the gaming bars, of course.

Management in gaming bars cares about one thing from their bartenders: making their "drop," the amount of money lost by customers during any given bartender's shift. Thus, it becomes the bartender's personal mission to try to keep as many people playing at all times. They accomplish this by dishing out free drinks and – if you're a regular with heavy losses and a penchant for big tips – food. Bartenders also give preferential service to gamblers because, if a customer hits something, they expect to get tipped well. This is what keeps Vegas running: the blood gutters from the slot machines dripping the hard-earned cash of true-blue losers into the managers' pockets. Creepy, huh?

At any rate, it's a simple system to exploit. The fundamental difficulty that must be overcome is one of mere perception. Well, perception and longevity. Basically, you want the bartender to think that you're there to play for as long as possible when, in fact, you're just there to drink for free for as long as possible. Here's an outline of a reasonably generic strategy (with a few special tips thrown in):

What You Will Need To be a total cheapskate, you'll need: a $20 bill, a $10 bill, and two $5 bills. With a little luck, we should be able to parlay this paltry sum (which would only buy you four shots chased with four large draft beers in most places … tip not included) into 12ish drinks or more (a fine start for an evening of getting wasted). Find a video-poker machine!

Your First Free Drink will always cost $20. As soon as you sit down at a machine, wait for a bartender, and (as you order) feed your money to the video-poker machine (always located on the bar) within eyeshot of the bartender (they can't comp what they don't see). When they bring your drink, say "thanks." They might say "good luck" or something to that effect, indicating that your drink is on the house. If they ask if you want to start a tab, you've probably landed a bartender (or place) that doesn't comp. In that case, abort. It's always a good idea to tip a couple of bucks after the first drink (as a sign of good faith), but it's not mandatory.

Play Slowly Gamble deliberately, but don't deliberate. Just be casual. Play max credits (on poker only) whenever the bartender is near you. You want to parlay your initial investment into as much time on the machine as possible. After your first 10-spot is gone, it's perhaps advisable for you to switch to a slower game (ideally, keno—tough odds, of course, but slow). If you're not asked to pay for your second drink, it's a good sign that they understand. Tip after they bring it to ensure speedy service.

Cashing Out Cash out only if you've at least doubled your money. In most places these days, even a small-stakes win requires the bartender to hand-pay you after you cash out. This means they'll know how much money they're making (or losing) from comping your drinks. If you bail on a machine for only $30, the bartender will know that you're not really there to play (which means he'll assume that you're there to pay). Therefore, my personal rule is: if you must cash out, do it only if you're sure the machine hates you and only bail when you've doubled your first investment. Tip $10 to the bartender, then move to a new machine and throw the $20 back in. If you doubled your money on the last machine, you should now only be down by $10 (roughly the cost of your two drinks), but have tipped the bartender very well (which is always good). Once again, play slowly and hope the comps keep coming.

The Goal Ultimately, it's to play for the better part of an hour off as little money as possible. After that, chances are that you won't have to pay for most of your drinks. When you lose (which is almost a certainty based on the amount of money we're playing with), feed the $10 to the machine when the bartender is away from you. Unless they are in front of you, they usually don't care about the denomination (apart from the initial $20). If you get lucky, you might be able to wage a comeback that'll keep you going for another 5 to 10 minutes. If not, then at least (from the bartender's perspective) it looked like you tried to recoup your losses with another $20.

Remember, after the first $20, you have to be in all the way for the full $40; the only reason you brought the second $20 to the bar in broken bills was to create the perception that you were feeding more money into the machine. Bartenders may like winners, but managers like losers and (in a lot of places) take care of them so they'll come back. Keep playing, but save the last $5 for the final step.

When the Money Runs Out All right, now for the moment of truth. Once you are down to the second $5 bill, stop playing and sit there and drink for a couple of minutes. If you want, you can move to a new machine that might make you look like you're considering gambling more. When you're ready, order another drink. This will be the telling blow: if you've managed to play for over half an hour, it should be comped. If it is, then the odds are good that the bartender won't hit you up for cash for the rest of the night. Use the last fiver to tip the bartender and drink as you normally would. You'll know that you've successfully scammed them if you manage to drink for the next two hours without being asked to pay.

If They Ask You to Pay Don't go nuts. Act surprised and say, "But I was playing … I'm in for $60 on that machine." Be nice when you deliver this line, don't yell, and don't use a whiney tone of voice; sound upbeat. Unless they're completely aware that you're trying to screw them, they'll buy it and should comp you. If they ask again on the next (or a later) round, unfortunately, you'll probably have to cough up some cash. At the *very* least, you should be able to get away with four more drinks (setting you flush with your gambling losses). Policies on comping customers who were (but are no longer) gambling vary from place to place, so (unless you've been there before) you can't be certain how far they'll let you go. That's also what makes it fun, and why we've invested in the bartender's good will with our tips.

One Word of Warning About Gambling This strategy is designed around the supposition that you will lose. The odds of winning anything are poor. If you like to gamble, use these suggestions as an outline. If you hate to gamble, you probably shouldn't use them at all. Either way, it's fun to see what you can get away with. Good luck!

The Best Game to Blow Your Last $20 On Everyone thinks that they'll strike it rich when drunk and in Vegas. Sadly, this is not the case. You're more likely to fuck Jenna Jameson during a random encounter than hit a Royal Flush on a video-poker machine.

That having been said, for the enthusiastic (yet inexperienced) gambler, the end of the evening often results in a difficult choice: how to lose the last $20 that you have on this earth. My answer is simple: blackjack.

Also known as desperation gambling, blackjack is a quick and easy way to find out if God loves or hates you. While there are plenty of actual books and theories on how to play, mine are simple: don't hit when your cards total 16 or more. Never buy insurance.

Always play four hands before quitting and never bet yourself dry before you've gone through four hands. And, finally, never play blackjack on a video-poker machine: the odds are insanely bad, and no strategy (beyond betting amounts that go well beyond the point of insanity) can save you.

And that's that. If you get lucky and manage to put together a good run, you can be back on top in mere moments. Or the cards can simply destroy you, consuming your pathetic offering in seconds. Either way, it's my personal pick for the best way to send that last little $20 into the great beyond.

Vegas Bar Etiquette

If you learned everything you know about Vegas from movies and TV, then everything you know is (very likely) wrong. That's OK. I'm here to help. Here are a few of the most common mistakes inexperienced visitors make in bars:

Not Tipping Tip your bartender/server. You will likely receive better service and (at the very least) will not be hated and reviled as a cheapskate. This is particularly important if you plan to return to the same bar for several nights in a row; bartenders have long memories here and never forget a face. In mid-range joints, the bigger the tip, the greater the likelihood that (over the course of several nights) you'll find large portions of your tab being conveniently forgotten by your bartender. If you gamble as well as tip, chances are you will be able to get away without paying for much, if anything at all. Of course, being a cheap bastard is your prerogative, but having the folks behind the bar as your friends (as opposed to enemies) is usually a good idea.

Cheering Loudly After Small-Stakes Wins Nothing says "clueless rube" like hollering "I WON, I WON! YEAAAHHH!!!" after you hit your first three-of-a-kind for $3. Save all acts of vocal, public rejoicing for wins that net you $200 or better and you'll find that those around you won't hate you quite as much. You'll also get better service by not screaming every time you're up by $20; bartenders expect to be tipped after you cash out, and sounds of victory draw their attention away from whatever they were doing before.

Physically Attacking a Gambling Unit Don't beat up a machine. No one cares that it took your last $20; chances are that the guy next to you is in much deeper than you can imagine. Don't play what you can't lose (and always assume that you will lose it). Remember: Vegas wasn't built on winners. And, yeah, behaving like this will likely get you arrested. If not the first time, then the next. You have been warned.

Complaining About Comps Always assume that you'll have to pay and *never* complain about not being comped for your play. Unless you've lost more than $100 at a place in one night, complaining will just make you look like an asshole who doesn't understand how the system works. This is doubly true of almost any bar in a casino. Do not demand to see the management and don't lie; cameras are everywhere and (worst-case scenario) they *will pull footage* of your play.

Spilling Drinks on Poker Machines Spilling a drink (or breaking a glass) isn't considered a big deal in most places; these things happen. However, spilling a drink on a poker machine is viewed as a sin against man and God alike. The reason is simple: poker machines are computers and computers get all fucked to shit when their innards are filled with an ice-cold gin and tonic. Malfunction voids all plays.

Puking Generally speaking, puking is fine, provided that you do it in a designated area and manage to spew on-target. Puking while standing at the bar is a big no-no, as is puking into a change cup (which happens more often than you would like to think). Puking in parking lots (of bars, not casinos) is also generally acceptable. Barfing at the Double Down Saloon will earn you a rag and a bucket and a sadistic bartender who'll be more than happy to introduce you to the house rule: you puke, you clean.

Falling Down Whatever you do, do *not* do this. Falling down is usually the kiss of death in a bar, demonstrating a complete and total lack of control over your own actions, proving to the world that you are, in fact, all fucked up. That having been said, it happens all the time, so (if you think you can stand—pun intended—the embarrassment) go for it.

Getting Cut-Off Getting cut-off in a bar happens once every 10,000 years here in Vegas, and usually only when one bartender at one place has fed you all the booze. That having been said, though, when someone does get cut-off, he usually deserves or needs it (regardless of whether or not he's aware of it). In nicer places, it's important for you to know that the bartender *will not actually say* that the brakes have been put on your drinking. Instead, they will often offer you water or some other non-alcoholic beverage instead of whatever it was you originally asked for. Be gracious and don't embarrass yourself by trying to argue about the bartender's judgment: you only get cut-off if you're far beyond drunk or if the bartender hates you. Either way, the results are still the same. Oh, yeah: usually, you'll be un-cut-off after you have a few waters, but (if you must drink right away) then head someplace else.

Getting 86ed The rarest and the worst thing that happens in a Vegas bar. Getting 86ed means that you have been such an irritating (or dangerous/threatening/violent) fucker that management has decided to extend a permanent invitation for you to go someplace else. The best advice is to get the fuck out of there as quickly as possible (preferably before the cops are called) and, whatever you do, *don't ever go back*. Regardless of whether or not your exile from the bar was justified, the choice between going someplace else and going to jail is a pretty simple one.

No joke: if management feels threatened (or pissed off enough) by your presence, the police will be called and you'll be arrested. Please, don't get 86ed (especially if you're holding a copy of this book).

And now that all the pearls of drinkerly wisdom in Vegas have been dispensed, let's have a few (too many). The following list contains almost all of my favorite bars in the blasted buffet of decadence that is the Las Vegas Valley. I've broken the places down by area, and (where possible) they're grouped together by proximity. That said, everything is a long walk in Vegas so don't blame me if you end up on a death march to Margaritaville.

THE BARS
University District

As mentioned earlier, the university district—the area surrounding the University of Nevada, Las Vegas (UNLV) on Maryland Parkway—houses most of my favorite dive bars. These are places where real people (and students) go. Generally speaking, these bars have a sense of character and history as old and as gritty as their clientele, offering an authentic Vegas experience tourists rarely see.

Our first stop is right across the street from UNLV. **Cheers Bar & Grill** (1220 E. Harmon at Maryland, 702-734-2454) is one of the town's best dives. The booze is cheap and the people are crazy. You'll meet an enormous variety of psychos here, ranging from college students and homeless people to used car salesmen and off-work limo drivers. It's the seedy nexus of Vegas, offering an almost statistically accurate sampling of the residents. It's also an excellent place to play pool; there are usually at least two serious shooters hanging out here. Be advised that they don't take credit cards and will not run you a tab. Be warned, the place has a tendency to get packed with hipsters on most nights during the semester. All things considered, though, if you want to see the real face of Vegas, look it up here; the place has enough character to crack your teeth.

Barfly note: you'll likely get one of three bartenders: Bernie, Radar, or Adam. They're all swell guys and (provided that you don't piss them off) will serve you well. Ask Radar to juggle (it makes him happy).

After you manage to get away from the crazy guy who likes to babble at random strangers about the spacecraft he's building, you should probably head a couple blocks south to the **Freakin' Frog Beer & Wine Café** (4700 S. Maryland Pkwy. just north of Tropicana Ave., 702-597-3237). The Frog has the largest selection of beer in the state, clocking in at more than 330 varieties (and still growing). It really is a world-class beer joint, even if it's a little on the small side (plans for expansion are already in the works). Very popular with frat boys (and their hot girlfriends), the place fills up with a young, clean-cut crowd after 9 p.m. during the semester. My advice? Hit the place early. Oh, and if you're worried that premium beer may be too pricey, they do keep staples like 40-oz. cans of Colt 45 and Pabst Blue Ribbon handy. They do not serve liquor. If heading there late at night, call ahead; the place closes after things slow down, usually between 1 and 2 a.m.

If you stood in front of the door to the Frog and looked to your left, you might notice the **Stake Out** (4800 S. Maryland Pkwy. just north of Tropicana, 702-798-8383) in the shopping center directly adjacent to the bar. The Stake Out is where acting students, playwrights, and theater techs hang out after rehearsals, but (much like Cheers) the place is another good dive. They have cheap booze and inexpensive food (which also happens to taste good). If you pay careful attention when you walk in, you might notice a staircase hidden by the main entrance. This leads to the second floor where there are pool tables and more drinking to do. Unfortunately, there's no service on the second floor so you'll have to shuttle your booze back and forth. Credit cards are accepted.

From the Stake Out, you could hang a left on Tropicana and head down to the **Crown & Anchor British Pub** (1350 E. Tropicana Ave. just east of

Maryland Pkwy., 702-739-8676), another perennial student favorite. The Crown is a British-style pub/sports bar with a lot of European beers on tap. If you're a soccer fan looking for the best place to watch the game with a bunch of screaming, like-minded individuals, then look no further. Almost perpetually slammed with students throughout the week, the place gets truly sick on the weekends. Avoid sitting at the tables if possible, as seeing your server will be a rare and fleeting thing. Table service here has been bad here for as long as I can remember.

There is also a **PT's Pub** (various locations) located in the shopping center at the corner of Maryland Pkwy. and Tropicana. **PT's** is a huge chain of bars in Vegas that offers a completely homogenized drinking experience. They're not bad, but they're considered far from interesting by the locals—think "McDonald's of Booze" and you'll get the picture. Still, if you're hoping to hook up with some ditzy co-ed cheerleader (or start a fight with her boyfriend), this particular location is probably your best shot, especially on weekends. Regardless, whatever you do, do *not* go to **Dotty's** (various locations); it is hell. With grandmas.

Starting from UNLV again, if you were to head north (i.e., away from Cheers) toward the corner at Flamingo and Maryland, you would come to the **Hookah Lounge** (4147 S. Maryland Pkwy. at Flamingo, adjoining **Paymon's Mediterranean Café**, 702-731-6030). The draw here is premium booze and specialty tobacco (*not* the wacky kind, unfortunately). You and your friends can sit around smoking any number of the house's exotic flavors using their state-of-the-art hookahs. The atmosphere is favored by locals, because it's a change of pace from the artificiality of the Strip and the dinginess of the dives, but expect to see some executive types here just the same. For non-smokers, there's always the bar at the back of the room. Try a double shot of Red Army Vodka; the bottle comes in a case shaped like an artillery shell and the "report" is just as subtle as it scalds your guts. The booze leans heavily toward the high end of the scale, but the place is worth checking out—even if it's for just one drink.

Hookahs are nice, but you probably came to Vegas with the need to lounge. If so, then you definitely must check out **Champagnes Café** (3557 S. Maryland Pkwy. at Dumont, across from the Boulevard Mall, 702-737-1699). There are few other places in town that embody the total essence of decaying old Vegas charm as much as Champagnes. Home to the elderly and the insane, its plush confines have served as a containment unit for many a tripped-out hipster's booze-addled brains. Cheap drinks, crazy atmosphere and comfortable seating. And new homeless friends! Does *not* take credit cards, but has an ATM.

If you're feeling hardcore, there's the **Huntridge Tavern** (1122 E. Charleston Blvd. at Maryland, 702-384-7377), an inimitably gritty dive that I heartily recommend. Small, compact, cramped, and psychotic, it's a wonderful place to tie one on. Also doubles as a liquor store: **Huntridge Package Liquor & Cocktails**.

Our last stop on ye olde College-Drunk Express takes us to one of the craziest fucking bars in the entire town: **Davey's Locker** (1149 Desert Inn Road at Maryland, 702-735-0001). Designed around a demented nautical theme, Davey's has been widely acknowledged to be the narrowest bar in Las Vegas. Tucked neatly between a Payday-Loan joint and a liquor store, the place is my personal pick for being the poster child of sleazy bars. If you're hardcore, then a visit to Davey's is a must; it will have you shouting "Dive! Dive! Dive!" all the way to the bottom of the briny deep.

Barfly tip: Hookers have been known to hang out here. We're not talking Julia Roberts in *Pretty Woman*, either; real grade-Z hoes. Caution is advised.

The Strip

A more wretched hive of scum and villainy has never been seen before. We must be cautious. But seriously, folks, somewhere between the traffic, the cops, the psychos, and the unbearable crush of tourists, the Strip can morph into a giant, vengeful robot, intent on scraping your mind of every last shred of sanity. In this section, I'll show you how this can actually be an enjoyable experience. All right, then. Let's hit the bars.

First and foremost, our tour begins at the **Peppermill Fireside Lounge** (2985 Las Vegas Blvd. S. at Convention Center Dr., 702-735-7635). In the '70s, the place was a posh supper club, now it is a world-class bar. Think of it as the less-seedy version of Champagnes Café and you'll get the picture. Put on a bowling shirt and your best *Swingers* impression and check it out. There's a good chance you'll catch anti-magicians Penn & Teller there by the cool-as-hell firepits discussing the finer points of sawing a woman's head clean off.

Next up is the **Laughing Jackalope Motel Bar & Grill** (3969 S. Las Vegas Blvd. at Mandalay Bay Dr., 702-739-1915). Located at the south end of the Strip (across from Mandalay Bay), this psychotic little bar's mascot is an anthropomorphic "jackalope" wearing a Hugh Hefner-style smoking jacket. Popular with service and casino-industry folks, the Jackalope features a full bar (they have Tecate on tap) and the largest hamburger I've ever seen. They call it the "Monster Burger," and it's as big as your fucking head (and, at $6, dirt-cheap). If you can manage to finish one, your next one is free. They also sell a "Double Monster Burger," but, as of this writing, no one has ever managed to consume one in its entirety.

Depending on how strict your budget is, you could then head across the street to **Mandalay Bay Resort & Casino**. Deep inside the bowels of this hotel (just ask the concierge) is **Red Square** (3950 Las Vegas Blvd. S. at Mandalay Bay Dr., 702-632-7777), the ultimate tribute to the Russian people's contribution to the world of drinking: vodka. They have every kind of vodka on earth, from the lousy stuff (Popov and Kamchatka) to things that you've never heard of nor imagined. Be warned, though: the place is pricey. A headless, birdshit-stained (not really) statue of Lenin out front adds immensely to the atmosphere.

For those who consider themselves to be night-hawks, then you should not miss the **Oba Lounge** at the **House of Blues** (702-632-7777). Open only on Friday nights starting at 3 a.m., the place becomes the venue for an after-hours DJ spin session. HOWEVER, the reason I'm recommending it has nothing to do with the music: the room is actually the VIP lounge at the House of Blues. This means that (after the obligatory 20-minute wait for service at the bar) you can take your beer and sit on the balcony overlooking the fully illuminated, yet completely empty stage at the HOB. It's the most unusual place I've ever drank at in Vegas, and I heartily recommend the experience to anyone who wants to keep the party going well past dawn. Open only on Saturday (early mornings). Cover varies.

If you like jazz, wander over to the **Bellagio combo lounges** (3600 Las Vegas Blvd. S., inside the Bellagio, 702-693-7111). You can find these mini-bar-lounge-hybrids scattered randomly throughout the hotel (just ask an official hotel employee if there are any combos playing and where to find them). The seating is comfortable, the atmosphere is casual, and the music is great (and free!). On most nights, you can find (at the very least) a piano-and-upright-bass duo playing here. The only caveat is that the booze doesn't come cheap. Think $7 screwdriver and you get the picture.

Another option for the budget-minded drinker/jazz aficionado is the **Baccarat Bar in the Mirage** (3400 Las Vegas Blvd. S. at Spring Mountain, 702-791-7111). Similar to the Bellagio only in that they offer free music without cover, the Baccarat Bar also has (slightly) cheaper drinks. Its close proximity to the insanely narrated (and completely free) white tigers on display by the Mirage Shops also makes it an entertaining place in which to get wasted.

For the unsubtly minded, there's always **Gilley's** (3120 Las Vegas Blvd. S. at Spring Mountain, inside the **New Frontier Hotel and Casino**, 702-794-8200). Lately, they've been pushing a lot of events on the general public, ranging from bikini bull-riding to mud-wrestling and wet t-shirt contests. It's the perfect place to unleash the psychotic redneck within us all … um, if you're into that … uh, kind of thing.

Heading toward the north end of Vegas Blvd., we

arrive at **Dino's Lounge** (1516 Las Vegas Blvd. S. at Wyoming, 702-382-3894), a popular dive bar frequented by only the most elite homeless denizens of this part of the Strip. Although I regard karaoke as a musical bloodsport, I have to say that their Saturday karaoke nights are some of the most insane things I've ever attended.

Our last stop on the Strip will also earn you a "Tom Waits/Charles Bukowski Citation for Daring While in Pursuit of Drunkenness" commemorative placard (see the entry on **Decatur Liquor** in the Karaoke section for instructions). By far the most frightening bar I've ever been to in Vegas, **Atomic Liquors** (917 Freemont St. between 9th and 10th, 702-384-7371) is … well, horrifying. I could tell a few stories, but then you wouldn't go there—and what's the fun in that? For the adventurous only. Disregard personal safety at the door. Be afraid. Be very, very afraid.

Barfly tip: I've heard reliable reports of whores sucking you off in the supply room out back. That's the kind of place Atomic Liquors is.

The **Artisan Hotel & Spa** (1501 W. Sahara Ave. at the I-15, 702-214-4000, www.theartisanhotel.com) makes it difficult to summon a vocabulary that would do justice to the lounge. Think deep mahogany. Think black granite. Think framed prints of famous works of art covering the walls and affixed psychotically to the ceiling. Yeah … the ceiling. It's covered in the stuff. Sitting at the bar makes you want to speak French, makes you want to date a Mucha girl or boy (whom you suspect might come to life and join you at any second). It helps you feel, for a brief and shining moment, that you've finally managed to return to the

1920s speakeasy, where you were *supposed* to have been born. Ephemera aside, the joint positively reeks of class. It's definitely one of the nicest, most tastefully chic bars in town. Oddly, the atmosphere isn't at all overtly cheeky or intimidating, just different and exceptionally nice. Too bad the drinks and food are INSANELY EXPENSIVE!

Off-Strip

Because there are some places within close proximity to the Strip that I would highly recommend any visitor experience, we've chiseled these special picks off from the rest of the pack. For the most part, they're on Paradise (a long block east of Las Vegas Blvd).

Pink E's Fun Food & Spirits (3695 W. Flamingo Rd. at Arville, 702-252-4666) used to be a place that was, quite simply, unbelievable to behold. Once upon a time, all of their pool tables were covered in pink felt and just seeing this sea of rose-tinted billiards was something else. As it stands, the place is still great fun. An enormous warehouse of diversion, it's the only bar in town in which you can rent ping-pong paddles and play until you piss your pants. The bathroom walls also feature a charming assortment of pornography clipped from *Playboy*. Try not to get mesmerized while you drain your lizard. Pink E's has air hockey, pinball, darts, championship pool tables, and live music.

Barfly tip: Despite popular belief, ping-pong is the most absurdly difficult game ever devised when playing under the influence of alcoholic substances. Misfiring motor neurons, slow reaction speed, and discombobulated hand-eye coordination make playing ping-pong while intoxicated highly uncool-looking.

Cover varies, but always cheap ($3-5).

It's always three in the morning at the **Double Down Saloon** (4640 Paradise Rd., one block south of the Hard Rock, 702-791-5775). Depending on where your hipster sensibilities lie, the place is either as pretentious as the annual meeting of the Andy Warhol look-alike club or as unreserved as four drunks fighting after a pissing contest. The décor consists of blackness offset by graffiti. There are no doors on the bathrooms. Don't order the house drink Ass Juice: it's the squeegee-ed fruit of the bar rag. Kickass live music all the time.

Just a few feet away from the Double Down, the **Office Bar** (4608 Paradise Rd. at Naples, one block south of the Hard Rock, 702-737-7756) is another fun place to frolic. I like this place a lot, mainly because (apart from attracting a regular crowd of defectors from the Double Down) the bar looks like it was never completely finished after construction. Also doubles as a package-liquor store.

The **Hofbräuhaus** (4510 Paradise Rd. at Harmon, 702-454-4009) is absolutely nothing like its original Teutonic cousin. That's also what makes it fun. The beer is insanely pricey, but it's worth it just to drink from the enormous glass steins. Plus they fly the stuff in fresh from a brewery in Munich; it's as close as you'll come to fresh European beer being drawn off the tap. Full bar, but the prices are just as inflated for the liquor, so forget it. Best visited midweek around 9 p.m., when it's not as packed as usual.

And finally, we arrive at the **Hard Rock Hotel and Casino**, where the elite meet to see and be seen. Our destination here is the **Circle Bar** (inside the Hard Rock Hotel & Casino, 4455 Paradise Road at Harmon, 702-

693-5000), a place of infinite merriment and horror. On the weekends, this enormous bar (which is, in fact, shaped like a circle stuck into the center of the casino floor) bears a striking resemblance to the eighth level of Hell in Dante's *Inferno*. If you manage to get a seat, be sure not to lose it or you'll be standing faster than you can say, "Hey, baby, what's your sign?" If you're hoping to hook up with a member of the opposite sex, it's a good location to be on the prowl.

Barfly tip: for the suicidal, hit this place around 9 p.m. on the weekends when the crush of people is so intense that you feel like you're being squeezed in an enormous vice grip of flesh.

The **Gold Coast's bowling alley lounge** (inside the **Gold Coast** across from the Rio, 4000 W. Flamingo Rd. at Valley View, 702-367-4700) really does come close to imparting the sense of being in *The Big Lebowski*. The bar itself is very small and not designed to accommodate many boozehounds at once, obviously figuring on the high turnaround as bowlers come in and return to their game. This gives the place one huge advantage: service here is like white-hot lightning striking your brain. Even when the place is exceptionally packed (and it can be on league nights), you'll never go wanting at the brass rail for that next drink. There's a comfortable dimness to the Gold Coast's bowling lounge I've never quite found replicated in other hotel-casino bowling alleys. Plus, the age of the alley really enhances the sense that you're indeed in *Lebowski* territory.

For the uninitiated, **Quark's** (702-697-8725) is a replica of the restaurant and bar from the series "Deep Space Nine." It sits in the center of **Star Trek: The**

Experience, surrounded by countless merchandise stores packed with "Star Trek" souvenirs, as well as the entrance to the two virtual-reality attractions and an impressive display of memorabilia. The servers dress in "Star Trek" uniforms, and various employees dressed as Klingon warriors, Ferengi merchants, Starfleet officers and even Borg drones walk around and make "Star Trek"-style conversation with you. Drinks are spiked with neon colors or served in glass spheres packed with dry ice.

Central Las Vegas

There's a lot to the creamy center of Vegas. So much, in fact, that it's impossible to tell where it actually begins or ends. Just picture this area as a drunken smear across the map, because by the time you're done here, that's pretty much how you'll feel.

I've tried to break this section up into clusters of bars to allow for as much walking as possible, but some places are simply out of the way. If cabs are completely out of the question, then skip to the section devoted to "Karaoke Alley," where there are at least four dives within easy walking distance of each other. Otherwise, let the carnage resume.

But first: the whores! Yes, **Play It Again Sam** (4120 Spring Mountain Rd. at Wynn, 702-876-1550) may be a strip club, but its principle contribution to the overall quality of life in the Valley is its role as a strip club operating without a door fee while *still selling cheap booze*. Unlike other strip joints, the management at Sam's has graciously opted out of ripping their customers on the price of hooch, making it a great entertainment bargain for impoverished drunks and devotees of topless dancing. Shots will run you $5 (which is pretty average), and bottled beer is a mere $2. Further bucking the trend, they will comp you for your play on the machines (highly unusual in a strip joint). The girls are not bad for being … well, basically free, and the atmosphere is extremely comfortable. Like all Vegas strip clubs, the place gets obscenely crowded on the weekends, so try to check it out when it's nice and dead (you'll get your pick of the girls).

Barfly tip: A stripper is a stripper, *not* your new girlfriend. The girls here are hardcore, desperate grifters and know how to play the game. Be sure to remember this before forking over all your cash as you go ga-ga for a shifty chick in a G-string and go-go boots.

After that debauchery, you may want to head a mere half-block south down Spring Mountain to the **Don't Ask Lounge** (3939 Spring Mountain Rd. between Valley View and Wynn, 702-876-4114), a seedy little dive with a pinball machine and a pronounced lack of class. While unremarkable in its own right, the Don't Ask serves as an excellent location for either post-topless decompression or pre-pubic lubrication (aka getting as drunk as possible *before* heading to the strip club). The booze is cheaper here, and its close proximity to **Play It Again Sam** makes it a natural pick for those on a strenuously tight budget.

Provided that you've not lost your ability to hear (and/or see), it's a short cab ride or a horrendously long walk to **Skip's Gold Coin Saloon** (4680 S. Decatur Blvd. between Tropicana and Flamingo, 702-871-4551), a bar that serves as a potent reminder as to why people often bring their friends out when they drink. Simply put, no one is ever at Skip's—the place is dead as

Dickens 24/7. This is unfortunate, as the place is actually quite insane; it has the second longest bar that I've seen in town, has a dining area upholstered in hideous naugahyde, comes appointed with pool tables and a jukebox, and is just generally enormous and strange. If traveling with a large party, it can be a lot of fun to take over the place. Also, its relative proximity to the Strip makes it a breath of sweaty, stale-tasting Vegas air, serving as a stark contrast to the pre-packaged (and packed) casino bars.

If Skip's sounds like way too much fun, then you may want to head to **Norma Jeane's** (3650 S. Decatur Dr. at Twain, 702-223-2570), a beautifully appointed tribute bar devoted to Marilyn Monroe. Yes, you read that right: a Marilyn Monroe *tribute bar*. Only in Vegas would such an insane thing exist, and Norma Jeane's is not to be missed. The place's drink prices are flush with any dive bar in town, allowing you to drown any shame and misery in a tidal wave of booze. The place is a guaranteed good time.

While you're in the area, you may want to consider visiting **Pool Sharks** (directly next door to Norma Jean's, 702-222-1011), one of the valley's best pool halls. If not, then proceed directly in the opposite direction to "Karaoke Alley."

Karaoke Alley

Karaoke. The shame. The horror. The misfired screech notes of insane musical violence. Shudder with me, fellow travelers. Shudder.

Karaoke Alley is a conflagration of dive bars packed into the area around Sahara and Arville, all of which compete fiercely for the valley's burgeoning karaoke demographic. Yes, while live music may have dwindled to little more than a steady trickle, our karaoke scene has managed to flourish, spreading like cancer. On any given night, these bars are packed with average, ordinary people pretending they're rock stars destined for a big shot on "American Idol." If you've yet to acquire the special flavor of shame and degradation on our little liquor-soaked tour, then Karaoke Alley will (like a communion wafer) staple the sensation squarely on your tongue. With few exceptions, the bars on Karaoke Alley are nothing to write home about. They are, however, almost as cheap to drink at as any place in Vegas and (depending on the night and the place) are usually pretty crowded.

The Crow's Nest (3805 W. Sahara Ave. at Valley View, 702-871-4952) is a very delicate blend of crappy dive bar and hardcore-karaoke hellhole all wrapped up in the atmosphere of a pseudo-punk rock club. This is actually one of the better and more interesting bars on Karaoke Alley; the vibe is exceptionally strange, the service friendly, and the booze dirt-cheap. A word of warning about the place, though: The Crow's Nest attracts an older, more gentrified brand of karaoke singer, and they take their "craft" pretty seriously. Hell, I once interviewed a guy there who described himself as a "karaoke artist." (I know, I know … words fail me.) Just something to remember before engaging anyone in conversation.

From the Crow's Nest, it's a mere "Irish stagger" to **Beefy's** (4601 W. Sahara Ave. between Decatur and Arville, 702-221-4266), a bar that proves, once and for all, that one *is* the loneliest number. I almost can't bring myself to recommend the place, simply because it's such a hardcore dive. The last time I was there, an utterly smashed woman offered to pay me $40 if I would drive her to pick up her mom. While that's not exactly a fair characterization of the patrons here (they're just neighborhood drunks), it does give you some idea of what to expect. The place is decorated in a two-tone theme of dust (old and new) that coats the blasted, Mom's-Diner-after-the-apocalypse atmosphere. The booze is cheap, the drunks are crazy, and (if you think you're up for it) your evening here will, at least, provide you with a story for your friends and neighbors back home. Oh, yeah: don't ask why they call it Beefy's, OK? You *don't* want to know.

Heading back towards the Crow's Nest, you'll eventually arrive at **Woody's Bar and Grill** (3101 W. Sahara Ave. just east of Valley View, 702-257-9663), which is probably the most normal bar on Karaoke Alley. Attracting a younger (and generally cuter) crowd, the brightly lit bar does a bustling business on karaoke nights. Because of this, service can be a little slow but it's definitely worth it if you're looking for a good place to try out your sleazy come-on lines on attractive, off-work waitresses.

It's a hop, skip and a terrible, inebriated fall to **Jose Hog's** (3190 W. Sahara Ave. at Spanish Oaks, 702-253-6013), a bar whose mascot is an anthropomorphic Mexican gangster pig who smokes cigars and wears stingy-brim fedoras. If it were a joke, the punch line would go here. In any case, the place has the largest selection of tequila that I've ever seen (more than 20 varieties) and has a penchant for hiring cute bartenders. Unfortunately, as of the time of this writing, there is no legend of Jose Hog.

While **King Tut's Pub and Poker** (6138 W. Charleston Ave. at Arville, 702-258-6344) ain't exactly *on* Karaoke Alley, it's pretty damn close. Decorated in an all-pervasive shade of anger-red, KT's is a nice place to go for shots and keno. The place sports a decent juke-box, pool tables, dartboards, and all the other usual bar amenities. I enjoy drinking there, though, because of its close proximity to a mental institution. While I've never actually seen any crazier-than-average psychos there, a guy can still dream, right?

Congratulations: you've managed to survive Karaoke Alley! To celebrate, allow me to buy you a shot at a bar that is sure to do you in. Welcome to **Decatur Package Liquor & Cocktails** (546 S. Decatur Blvd. at Alta, 702-870-2522). This large, angry, neighborhood bar and package-liquor store positively redefines the word "seedy." If you actually end up drinking here, you're eligible to receive the highly coveted "Tom Waits/Charles Bukowski Citation for Daring While in Pursuit of Drunkenness" memorial placard. Simply send a photo of yourself at the bar accompanied by seven creeping Charlies to: Seth "Fingers" Flynn Barkan, 4075 S. Durango Dr., Ste 111-72, Las Vegas, NV 89147.

If you were to head west down Charleston, you'd eventually run into the **Tap House** (5589 W. Charleston Blvd. just east of Jones, 702-870-2111), a sanctuary of booze the likes of which few have seen sober. While not much more than just another neighborhood bar, the place reeks of authentic grit and madness. Don't be scared, though; the place isn't as dangerous as others I can think of. Reeks of "old-Vegas charm." Also serves reasonably priced food.

Continuing the long, boozed-out trek down Charleston, we'd arrive at **Frankie's Bar & Cocktail Lounge** (1712 W. Charleston Blvd. at Shadow Lane, 702-387-9256), a most unusual little bar where you're waited on by friendly, elderly matrons. Essentially a small square with a booze-filled creamy center, this place is almost always devoid of human life. (I can't speak for insect or animal life, but I have my suspicions.) It's a good place for a quiet drink or to get blasted through the floorboards.

Barfly tip: if you happen to be visiting Vegas during Christmas, then you have to visit Frankie's; they deck the bar with boughs of festive bullshit, transforming it into a Yule tidal wave of drunken madness.

And now we must journey to the east … the Far East … of town.

East Las Vegas

Ah, East Las Vegas, home to Green Valley, freeways, and SUV-driving soccer moms. Because most everything is out of the way out here, I'll just try to hit a few of the high notes, hotspots, and personal picks for places in which to get blasted. One word of warning: you'll be taking lots of cabs, so choose your places wisely. Nothing is close to anything in Green Valley.

First, join me in a moment of prayer as we enter hallowed ground. Due to its relative proximity to the university district, **T-Bird's Lounge** (9465 S. Eastern Ave. just south of I-215, 702-361-6639) has traditionally attracted small groups of underage teenyboppers (at one time, myself among them). These days, they card rigorously, although it's nice to remember "when." Still a personal favorite among the more divergent (psychotic) hipsters, the place features cheap food, cheap booze, and a homey (dive-bar) atmosphere. Welcome to Green Valley.

Meanwhile, on the opposite end of Green Valley, there's a nice little restaurant and bar known as **Henry's American Grill** (237 N. Stephanie St. at American Pacific, 702-898-5100). If you get tired of shitholes (but don't want to actually have to pay for an upgraded atmosphere), then Henry's has you covered. For reasons that are beyond me, the small bar that adjoins the restaurant draws practically every off-work bartender from the area. It's a good place for spirited conversations with random strangers. Small, dark and enjoyable.

Note: be sure to check out the insane wall of caricatures (aka "The Wall of the Damned"). These are the faces of poor bastards who blew enough cash on the machines until they eventually hit a royal flush. As punishment for their acts of defiance, their souls have been removed and entombed in these ridiculous renderings of their faces, trapped for all time, smiling insanely at the bar as a warning to those who would share their fate. Neat, huh?

If you were to take Las Vegas Blvd. South almost until you hit Primm, Nev., it might be possible for you to find the **Hurricane Bar and Grill** (10420 S. Bermuda Rd., just south of Silverado Ranch, 702-407-8976), one of the most improbably located booze halls in town. Truly stuck onto the ass-end of nowhere, the place serves almost no discernable purpose apart from their jazz-and-blues jam sessions. The décor consists of a bizarre mélange of sports-related memorabilia commingled with a random assortment of nautical/Caribbean-themed items. They also occasionally book decent music acts to try to draw a crowd, so call if you're in the area and are thinking of stopping by—you'll need directions. Note: At this point, I invite you all to sing with me: "I want a girl ... [hick] who will laughferno ... [hick] ... one else. ..."

Plunging into the center of the bucolic little suburb, we come to **Murphy's Pub** (3985 E. Sunset Rd. at Annie Oakley, 702-458-5516). I love this place; if your significant other has any capacity to enjoy a great dive bar, then I heartily recommend Murphy's. It's a large, rose-colored bar with a small stage and dark tables (perfect for drunken pawing and slobbering in public). They usually feature live music on Wednesdays and often run the show until after midnight. Cash only, no credit. Does *not* run tabs.

Barfly tip: If you end up leaving late at night, do not be surprised if you hear someone shrieking as if for help across the street from the parking lot. One of the nearby homes houses a menagerie of exotic animals and (when drunk) their caterwauling is often confused for human cries of distress. No kidding.

The **Ballpark Lounge** (7380 S. Eastern Ave. at Warm Springs, 702-361-1961) is a weird place across the street from Sunset Park. There's not much to do here but drink and nothing really special in the way of atmosphere. The bartenders are friendly, though, and (because it's very often dead) are usually more than happy to comp generously if you gamble.

GW's Place (3342 S. Sandhill Rd. at Desert Inn, 702-454-6100) is a strange place. Once this bar was compared to an immense tan straightjacket. It feels like a padded cell, covered in puffy tan "pleather" offset by sterile-feeling mirrored surfaces. Don't let the unusual décor fool you, though: it's a grade-A dive, replete with pool tables, inexpensive drinks, and (of course) madness. Lots of madness.

And finally, we arrive at **Chilly Palmer's** (1640 W. Warm Springs Rd. at Arroyo Grande, 702-456-2520), a bar whose mascot (of the same name) dresses like a turn-of-the-century Dublin dandy. CP's is a bit classier than some of the other bars on this list, but (for reasons that are beyond me) that doesn't seem to affect the drink prices. The shelves behind the bar are all made of nice dark wood, which help to give the place a faux-snooty undercurrent. They also have reasonably cheap food and a comfortable dining area. Be wary of some of the regulars, though, as they'll chew your ear off with their insanity.

Westside

The last area we'll explore on our toxic foray into total drunken oblivion is also one of the craziest. The Westside of Vegas is where we house our nicer, off-Strip drinking establishments. I'd be surprised if you managed to make it back to your hotel after an evening out here.

The **Kopper Keg West** (2557 S. Rainbow Ave. at Sahara, 702-254-5495) has been in its current location for the majority of my life. Stepping into the place is like entering a time warp into the 1980s. Naughahyde, mirrors, and dim, pink lighting enhance this sensation. If you're looking for a bar in which to try out your Tom Waits impression, this is it.

The **7-11 Bar** (2520 Arville Rd. at Sahara, 702-362-2415) has absolutely nothing to do with the convenience store of the same name. It's a very dark spiderhole of authentic old-Vegas atmosphere. Obsidian tables, pink neon, and patrons who like their bars quiet and their drinks cold make the 7-11 one of the most unusual bars I've been to in Vegas. Highly recommended.

Big Dogs (6390 W. Sahara Ave. at Torrey Pines, 702-876-3647) is an immense circus of canine-themed lunacy. Best if visited late at night (when the place empties out), they have several microbrews on tap and a solid menu of bar favorites. Draws large crowds of thirtysomethings from the surrounding neighborhoods.

And now, for the serious drinkers, I welcome you to experience a segment of our fair city that is (in the words of the late, great Wesley Willis) virtually guaranteed to fuck you up like a car crash. Yes, it's time for Bar Row.

Bar Row

Durango Drive is a long street that runs north and south, crossing over most of the main drags along the way. The south end of it has a whole slew of bars within walking distance of each other, running the gamut from corporate sleaze pits to Irish Pubs to pseudo-dives. This is one of the fastest growing areas of town; three years ago, there was hardly anything out here.

Because of the rapid pace of development, the bars out here are relatively new and are generally nicer than those in other areas of town. However, due to the fact that they are locked into close-quarters competition with each other, the prices are on par with most other bars. These factors have led me to call this area Bar Row, one of the few places in town where you can walk to a bunch of bars in one evening without the risk of sunstroke.

I'll present the places in order of preference, but I think it would be useful to note that, in geographical order (starting from the south end of Durango and moving north), you could take a cab to Bootleg Benny's and then hit all of the following places as you walk north: Chuy's, Putter's, Sean Patrick's, Timbers, (PT's, if you wanted to), Brewske's, and La Louisianne. Here's what you can expect:

Winner of the Most Poorly Implemented Theme Award, **Brewske's** (3645 S. Durango Dr. at Twain, 702-242-8554) is a tepid little hole that I regard as my home away from home. Located directly across from Desert Breeze Park, the place features three main qualities: super hot bartenders, generous comps, and all the booze you can drink. The service is also excellent. Attracting a diverse crowd of area drunks, off-work service-industry people, and late-night club chicks, you never know what you'll find here. Has three pool tables. Accepts credit cards. *Barfly note:* I've slept here. Once.

Bootleg Benny's (4705 S. Durango Dr. in between Tropicana and Flamingo, 702-450-4705) is an enormous barn of booze lined with $20,000 worth of flat-screen plasma TVs that are set (in perpetuity) to ESPN. The crowd here varies, but much like the other bars this far out west, the clientele runs heavily towards upper-middle-class white kids in their mid-20s. The atmosphere is schizophrenic (techno 1920s mobster neo-Vegas), but that's why God invented liquor. Comes complete with cheap food ($7 pizzas). *Barfly tip:* If you run into a hot Mormon bartender here, tell her Seth misses her. Also: Check it out during the graveyard shift for Nick, one of the most entertaining bartenders in town.

Jackson's (6020 W. Flamingo Ave. at Jones, 702-362-2116) is a strange, presidential-themed bar chain that's trying (in vain) to compete with the endless army of PT's Pubs that have popped up on the corner of every major intersection in town. This particular location is an exception to my usual hatred of corporate bars in that it attracts a fair amount of club girls (and their boyfriends) on most nights; the scenery can be nice. Also serves excellent food. Twenty different burgers on the menu!

Chuy's Mesquite Broiler (4460 S. Durango Dr. at Peace Way, between Tropicana and Flamingo, 702-873-7732) is a family-oriented trip into booze-induced dementia. If you can hit the place a couple of hours before they close (midnight), though, you'll be able to dodge the screaming children and pissed-off parents. This would allow you to enjoy the place's unique, beach-themed atmosphere in relative peace. While they do have a reasonably stocked bar, the best thing about Chuy's is their ultra-cheap, gut-busting chicken-and-cheese quesadilla. Just half of one will you give you more than enough ammunition when you vomit.

I'll be honest: I hate **Sean Patrick's Pub** (8255 W. Flamingo Rd. just west of Cimarron, 702-227-9793). It's an "authentic Irish pub," an "innovation" that began creeping onto the Vegas scene in the mid- to late-'90s. I hate them. I hate them all. We don't co-opt the drinking styles of the Vietnamese or, oh, say, the Australians, so why do bar owners feel the need to rip off the Irish? The place is there, though, and (although I hate to admit it) I do like the small booths—perfect for intimate, public make-out sessions. Oh, yeah: it's on Flamingo, one block east of Durango. Check it out if you have a fetish for "authenticity." Pass on it otherwise.

Finally, our last stop on Bar Row is **La Louisianne** (3655 S. Durango Dr. between Twain and Spring Mountain and behind Brewske's, 702-869-3154). It's a Cajun-themed restaurant that books authentic musical acts (amongst other, less-authentic bands, DJs, rappers, etc.). In a bizarre twist, they've started hiring private belly dancers who'll give you table dances on Friday nights. Check it out: it's very strange.

Still haven't had enough yet? Fine. Welcome to the **Roadrunner Grand Canyon** (9820 W. Flamingo Rd. at Grand Canyon right off the I-215, 702-243-5329), the veritable Death Star of bars. The service here sucks, but that's because it has the longest bar I've ever seen anywhere on earth. An immense, steel-plated monstrosity, the bar extends across the entire width of the place, almost managing to hit a vanishing point at the far end. They have mini-bowling. They have a menu that reads as if written by a rich hillbilly while she chased handfuls of Prozac with swigs from a bottle of Scotch. The place, simply put, is just not right. Should be seen once by all visitors to Vegas.

Finally, for those who simply can't be impressed, I have one last offering in this, our tour through the gutters of Vegas. Ladies and drunks, I give you **Sedona** (9580 W. Flamingo Rd. at Fort Apache right off the I-215, 702-320-4700), a rather expensive, very nice, monument to grade-A Vegas class. If this place doesn't make you go "oh, wow," then nothing will. Sleek glass surfaces, niche lighting, useless space, and a thoroughly Californian design concept combine to transport wealthy lushes off to dreamland in this black-turtlenecked concept bar. Be sure to visit the outside areas that allow you to check out the stars. Oh, and bring your wallet. The place is expensive.

STRIP JOINTS, BORDELLOS, AND SEX CLUBS... OH MY!

While sex is generally a source of commerce in most cities across the nation, few compare to Las Vegas for the wide spectrum of possibility from tease to release. Still, in a city that prides itself on freedom and decadence, there *is* a twist. What may be suggested may not be what's actually offered, and what's offered may be surprisingly different from what might be expected. Add into the mix the fact that law enforcement, variably interested in the going-ons in most places of adult orientation, has a direct impact on what really happens in this ever-changing landscape. The basic truth—that with enough money anyone can come to Las Vegas and get off in any way they like—holds, but it might take more effort than following the ubiquitous sexual advertisements adorning cab tops, billboards, aggressively marketed glossy flyers, yellow pages, and newspaper ads. In the pursuit of sexual satisfaction, the following will be indispensable.

Nightclubs

The ultra-lounge trend hit Las Vegas a few years back like a big bag of body glitter to the head. Every hotel-casino seems to have one. To draw the necessary crowds to get the buzz, these clubs almost universally rely on ad campaigns that show one or more impossibly beautiful models about to get it on with: A) a man in a shiny shirt; B) the other impossibly beautiful model in the ad; or C) both. So charged with the confidence that all a shlub needs to do is don some gay apparel to get it on in these nightclubs, he trots off to provocatively named places like **Tabú** (inside the MGM Grand Hotel and Casino, 3799 Las Vegas Blvd. S. at Tropicana, 702-891-7183) or **Risque** (inside the Paris Las Vegas, 3655 Las Vegas Blvd. S. just north of Tropicana, 702-946-4589) only to quickly face the reality—starting with the velvet-roped snobbery made popular in 1970s New York discos. Still, patience is a virtue, and eventually the doors to the kingdom will open up (and with it your wallet for the high cover charge). Of course, unless you agree to purchase a bottle of liquor (starting price typically around $200), you don't even have a seat inside. Soon it becomes clear real sex is *so* not part of the program. Indeed, as these clubs exist within regulated casinos, the Nevada gaming authorities have seen to it that no *actual* nudity or sexual activity takes place on the property.

While the early days of the club scene boasted legendary private areas and an atmosphere electrified with public sex, million-dollar fines and threats to gaming licenses have placed a definitive stop to all that. Not to say that the intriguing almond-eyed beauty from Senegal in the micro-tiny Bebe top is adverse to deep, passionate kisses as the techno thumps from speakers embedded in the multicolor lit tiles. After all, these clubs are filled with lots of horny people attempting to make Las Vegas their private *Boogie Nights*. It's just that if you're lucky enough to hook up (and still have some money left in your pocket), you need to get a room— and that's if she's not a "working girl" (see below).

Strip Clubs

With the large number of large strip clubs that have opened in Vegas over the past five years and only a few of the smaller clubs folding in the interim, it's safe to say that this town is at the peak of the Strip Club Era.

At any given time, there are more than 10,000 erotic dancers licensed to do their trade within the county limits. There was a time when it was common to see "everything goes" in the VIP rooms of most of the well-known clubs and in the public areas of the smaller ones. It was never a given that it would happen, and the entire transaction was done with at least a veneer of subtlety, but with the right dancer and denomination it was a fairly safe bet. Most of that changed in 2003 with a couple of FBI investigations cheekily referred to in the local press as "Operation G-sting." All of a sudden, an emboldened local vice squad stepped up citations, and business owners were seeing potential profits flutter away with one fell swoop of the county-commission gavel.

As a result, self-regulation has been increased, and the days of naughty dancers are virtually gone. Sure, most clubs still have the dirty girls with sexual-ninja moves who will continue as long as full-body contact is allowed, but don't expect a blowjob even after spending $1000 or more (all dancers will work you for maximum value) in the back room unless you hit one of the few remaining "working girls" still in the circuit.

Know, however, that most clubs have video-surveillance cameras everywhere. Not only can wrong moves trigger quick security responses and a toss to the curb, but it might be recorded for some ambiguously malignant purpose—just ask the city councilman who had a taped dance show up in an FBI search warrant return. That said, billions of "Vegas, baby" party men and horny loners continue to enjoy the sexual charge off the wide array of hard-bodied, fantasy-calendar fleshpots of dorm-room dreams. With the right moves, it's not hard to have a long interaction in the delusion that the dancer really, really likes you.

Following is a menu of noteworthy clubs. All the others are just variations on the theme. As a general rule, cab drivers will recommend the places that give them the largest fee for dropping you off, so take taxi suggestions with a grain of kick-backed salt. Totally nude places all charge a cover, and, with the one noted exception, only serve soft drinks (which are very expensive). Topless places are typically free in the morning and daytime and have a cover and a drink minimum in the evening. Women are welcome in most all clubs. However, with the exception of the male revue (listed below), there's the requirement that they be accompanied by a gentleman. Most places offer lap dances (unless otherwise noted) at $20 per song and have back rooms where there's either another cover charge or a minimum (unless otherwise noted) that starts at $100.

Can Can Room (3155 Industrial Rd. at Stardust, 702-737-1161, www.cancanroom.com) *Totally nude*. Still standing after all these years, the entrance routine has been honed to precision. At the door, there's a hard sell to pre-purchase a private dance that will cover the cover charge and give you unlimited soft drinks. With time-share-pitch intensity, you'll be offered a variety of options, including the two-girl show for an hour that will have a "pre-set" price of about $1,000.

Know that all these shows are negotiable, and a one-girl private-dance ticket for a full half-hour is available for $100 if you have a good gift of gab. Once the ticket is in your hand, you can peruse every dancer for your private show as they make an appearance on the large poled stage; there are no traditional lap-dance areas. With tasty Sprite beverage in hand, you pick your dancer(s) to go into one of five back cubby holes equipped with beds and chairs. (Some girls aren't available for the pre-paid ticket price, because there's a very Socialist system where every girl has to go once before any girl can go twice).

Generally, you can talk the women into the "bed show" for an additional $100. Then, for the next half-hour the naked woman will perform for you in ways that used to be called "second base" in junior high. The patron can stroke the non-naughty bits and writhe around with the action, but it can't get too sexual or trouble will ensue. Girls graciously take additional tips as the dance unfolds, but it won't get you any farther. At the *Can Can*, you Can't Can't. 7:30 p.m. until dawn. Minimum cover price: $20 plus drink minimum.

Cheetahs (2112 Western Ave. between Sahara and Oakey, 702-384-0074) *Topless*. Used to be known as the place to go for the down-and-dirty girls. That was, until the federal indictments were handed out. Traditional small strip club with a main stage, side stage and two VIP rooms. This is the place where Joe Esterhaus's classic skank-noir, *Showgirls*, was filmed. VIP room No. 1 consists of rows of benches where patrons are buttressed against one another in a virtual lap-dancing daisy chain. VIP room No. 2 is a little darker with individualized booths where the price is three songs for $100. Less private lap dances are also available on the floor.

Tip: During the early morning hours, say between 6 a.m. and noon, VIP room No. 2 drops to $20 per song and the floor dances drop to $10.

Indeed, Cheetahs is one of the few clubs that actually has action in the coffee-and-pastry part of the morning. Dancers are always friendly and frisky, but because of the screws coming down, it's just dancing with an occasional grope. Local radio stations sometimes do live broadcasts during the day and run drink specials. And sometimes they serve hot dogs. Mmmm, yum: strip-club hot dogs. 24 hours. Free before 6 p.m.; weekdays $5; weekends $10 cover charge.

Club Paradise (4416 Paradise Rd. directly across from the Hard Rock, 702-734-7990) *Topless.* The original classy joint, generally considered to offer the most traditionally beautiful dancers who oftentimes have actual dancing skills. Fairly large club filled with brass, marble and a gourmet kitchen serving up what's purported to be really good and expensive food.

Dancers are generally all business, and the grinding found in other clubs is a rarity here. Back areas are really for rock stars and rich folks, but tales of steak, lobster, champagne, and intense attention from a bevy of beauties are legendary. Still, the average Joe looking for sleaze is going to wind up broke and disappointed. For a virtually wholesome take on the whole concept, however, this is a club you can take your father to. 6 p.m.-6 a.m. $10 cover charge.

Crazy Horse, Too (2476 Industrial Rd. just west of the Strip and directly under the Sahara overpass, 702-382-8003) *Topless.* Another victim of federal search warrants and scrutiny, Crazy Horse, Too, had the reputation for wild times in the back room, but no more. Two stages in two separate rooms, it still seems like most the action occurs behind the black glass of VIP Land. Despite big-talk promises from the dancers, not much different from the full-contact dancing on the main floor, though. Either way, expect a steady assault directed at your credit card. Are you strong enough to resist?

Warning: there have been a recent rumors of assaults of the *physical* nature against patrons who get out of line or who complain about the final tally. No one's been arrested yet and no lawsuits have been filed against the club. All in all, it's a mid-level club that's easy to figure out upon entry with a very simple layout and mostly hot dancers giving what your fiancée at home probably expects your lap dances to be. Open 24 hours. Free before 6 p.m.; $20 cover charge.

Déjà Vu (3247 Industrial Rd. at Fashion Show, 702-894-4167, www.showgirl.com) *Totally nude.* As a totally nude club within the technical city limits of Las Vegas, this is one of the few remaining strip clubs that provides 18-year-old naked girls as an option.

Designed to lead you to believe that sex will happen if you just spend enough money, strippers are super-high pressure and will be soliciting back-room dances almost immediately. There have been a number of experiments over the years including "shower shows" (really wet girls who won't get you off), but the tried-and-true maze of cubbies with beds is the mainstay.

Back there, the object is to convince suckers that the more they spend, the more they'll get. Offers for this service are extremely pricey, and what you *do* get is indistinguishable from the $40 dances in the semi-private back booths. "Real privacy" is often the major selling point, but bouncers troll the hallways like school-district security guards with chips on their shoulders from not making the cut at the police academy.

In fact, unlike the Can Can down the street, there's scant body contact of any sort at any level. Essentially, a great place to see nubile, sweet youngsters showing you the parts of their body up close that you wish you saw when you were a daydreaming high-school nerd. 11 a.m.-4 a.m. (weekends till 6 a.m.). $10 cover charge.

Larry's Villa (2401 W. Bonanza Rd. at Rancho, 702-647-2713) *Topless*. Larry's Villa is the granddaddy of far-off-the-Strip strip clubs in Las Vegas. It's quite possibly the skankiest, most wonderfully horrible, low-rent nudie joint you could ever hope to get shitty on Johnny Walker Black in. The decor features large signs in bold print warning users and sellers that they will be reported to the police for drug activity and run-down dancers well beyond their prime, or younger ones just a little too odd to fit into the more mainstream clubs.

Also, a rough-and-tumble clientele (though generally harmless) who are more interested in talking up dancers than paying attention when they're actually dancing. It's kind of like an extremely dysfunctional dating service, and by most indications, there's a bond between customer and dancer like liquid cement. There are nights when cigarettes dangle from mouths of dancers on the '70s-style runway (which is probably when Larry's first opened), not to mention the occasional passed-out patron and/or dancer next to another passed-out patron and/or dancer. Larry's has a back room, but really: why bother? It's filled with the most uncomfortable-looking chairs and a giant camera, the monitor of which can be viewed by basically everyone else in the club who might bother to glance over.

Still, if you're looking for a place to engage in genuine conversation with a woman who gets naked for a living before going home to her humble apartment or trailer, there is, and will never be, a place better than Larry's Villa. Highly recommended, especially for people interested in writing short stories. 1 p.m.-5 a.m. No cover.

Las Vegas Lounge (900 E. Karen. Ave. behind the Las Vegas Hilton, 702-737-9350) *Topless trannies.* Las Vegas's only transsexual bar and surreal vortex. The clientele is a hodgepodge of gender and orientation with the only commonality being the love of pre- and post-op trannies prancing about in stripper wear.

Not that there isn't some reason to the rhythm; allegedly a diva lipsynch show is going on. These divas swing to their own beat and make it, literally, only kind of a drag. They might start on the official stage, but soon are trolling the floor for all-too-willing patrons to line their bodies with tips. Nipples are obscured by opaque cover or maybe electrical-tape Xs. Table dancing is offered with no "official" touching. There's always a cruise vibe in the air—well, boys will be girls. No cover but sometimes a drink minimum. Shows start at variable times, but usually around 9 or 10 p.m.

Olympic Gardens. (1531 Las Vegas Blvd. S. at Oakey, 702-385-8987, www.ogcabaret.com) *Topless.* In the early 1990s, this one-time Greek restaurant morphed into one of the most popular strip-clubs in town. Why? Well, it wasn't because the OG had any renovations. The club is aging like a grand dame showgirl after the makeup comes off. Still, with one of the highest dancer-to-customer ratios, what's a little duct tape and ripped-up carpet?

The back room is nice and secluded and probably looks like the VIP spaces at those fancy nightclubs that you never get to see with the bonus guarantee of nudity, but little else. For the ladies only, the OG offers the "Men of Olympus" male revue in an upstairs room where bodyhairless men in leather flop jocks tempt the ladies to scream like a Jerry Springer audience until "Apache," "T.N.T." and "Tiki" jiggle it, just a little bit. ("Hi, I'm the Construction Worker. Do you want to touch my Hard Hat?") Catering mainly to bachelorette parties, it might be weird to see the ladies beeline through the main floor. Or maybe it's weirder to see them an hour later, worn-out and stumbling outside. Where are they going?

Well, the smart ones wander across the street to **Dino's** where the karaoke barfly scene is the perfect antidote to too much oiled-up bulging. Then, there's always **Luv-It Frozen Custard** in the south parking lot. Cheap booze or dessert binging? Either way, it's the perfect end to the imperfect sexual encounter—just like in real life. Open 24 hours. Free before 6 p.m. with one-drink minimum; $20 cover (for the male revue, too).

Palomino (1848 Las Vegas Blvd. N. between Owens and Lake Mead, across from Jerry's Nugget, 702-642-2984, www.palomino-club.com) *Fully nude and alcohol.* The only club near town that allows what's obviously a dangerous combination of alcohol and female genitalia, the Palomino is as close to old-school burlesque house as you'll find in Vegas.

As the evening progresses, the main showroom opens. Dancers are fully adorned with festive costumes. Sometimes they even have a comic/MC running the show with expected hack jokes and shtick. Earlier in the day, however, all the action is upstairs in the mini-clubs with tiny stages and staging areas where private attention can get expensive.

Located in a funky, older part of town, the trip may appear to be a touch sketchy, but the neighborhood isn't as bad as it looks. Plus, just down the street, there's an ancient bowling alley at the **Silver Nugget** that might complete the delusion that you've been transported to a different time when strippers cared about presentation, all bowlers had crewcuts, and the hookers were the only ones wearing blue eye shadow. It may not have been a better time, but you always knew where you stood. 5 p.m.-5 a.m. $30 cover charge.

Sapphires (3025 S. Industrial Rd. at Desert Inn, 702-796-6000) *Topless.* Notable only for the fact that this has to be the largest strip club in the known universe. Jay Leno once referred to it as "Hooters meets Costco." The owners transformed the state's largest athletic club into what has been touted as the new generation of mega-strip clubs.

Opening night saw a performance by Motley Crüe and the elite politicians, movers and shakers of Las Vegas mingling with the horny and curious. Quite a party, indeed. The sheer magnitude of the joint actually makes all efforts at appreciating the stripping on the main stages impossible. Plus, the stages themselves are so awkwardly designed in an attempt to be "hip" that the dancers can barely move around without threat of serious injury (especially as they're all perched on the highest of heels). If you can, get a tour of the facility from one of the dancers. It's a neverending journey through lush side rooms, back rooms and sky boxes, each equipped with their own special allure. With prices rolling into the thousands, one can only wonder what happens if enough cash is doled out. Open 24 hours. $20 cover charge.

Sherri's Cabaret (2580 S. Highland Dr. at Sahara, 702-792-9330, www.sheriscabaret.com) *Fully nude.* Sherri's Cabaret is owned by the same people who run a legal brothel about an hour outside of Las Vegas. Unlike its bigger, more lavish sister property to the west, Sherri's is a small dive with a tiny "bar" downstairs and an equally small stage upstairs that encourages voyeurs to check out all the goodies of a completely naked lady up close and intimately. There's a one-drink minimum, and a bottle of water costs $8. Change comes back in two-dollar bills.

The lap dances are also up close and fully nude, and while they have a good deal of grind, there's nothing too overt happening. Then there are the "den dances." Starting at $175 for 15 minutes, bumping up to $225 for 30 minutes and on, you are escorted into a small room with not quite see-through curtains. The curtains are closed and tied closely together. A tip is negotiated with the dancer along with frank talk of what will and won't occur. At the end of the purchased time period, the cocktail waitress will gently remind the occupants by speaking through the curtain that "time is up." Dens tend to get much repeat business. Open 24 hours. $20 cover with one-drink minimum after 6 p.m.

Talk of the Town (1238 Las Vegas Blvd. S. at Park Paseo, 702-385-1800) *Fully nude.* More of an adult bookstore with a strip club in the back. This joint only recently dropped the motto, "Home of the $5 Lap Dance." The tunes come from a jukebox that shuts off (mid-song, if necessary) at precisely three minutes, which runs $20. This place is a real makeshift sleaze pit that doesn't aspire to do much more than make you wonder if you've landed on the Island of Misfit Strippers.

There are three rooms designed for intimate dances at affordable prices, but it's just more general weirdness. They also have a closed booth where the patron goes on one side of a Plexiglas, the dancer on the other, and a steamy conversation ensues where no one stops the patron from full expression of all the range of emotions and feelings that must build up talking to a naked girl through a Plexiglas window. Bookstore open 24 hours. Dancing 6 p.m.-4 a.m. $12 cover charge.

Treasures (2801 Westwood Dr. at Highland and Sahara, 702-257-3030) *Topless.* A more accessible Club Paradise. Ornate design, lush setting and even quality food service have made this the instant favorite of middling celebrities and men who like taking their wives to a classy nudie bar.

Of course, trouble brews deep in the caldron of flesh trade. Barely a year after it opened, law enforcement began a series of stings resulting in a number of citations for things getting too steamy and solicitations of the "oh really?" variety. Not wanting to lose their license, the club implemented a firm "no naughty" policy clearly posted across the tabletops. So no more alleged reports of Dennis Rodman getting his rocks off in the back booth, this place is strictly … pretty. Also, in an attempt to lure more daytime customers, the kitchen offers free lunch. Try the tuna melt. (No joke.) Open 24 hours. $20 cover charge.

Swingers Clubs

Red Rooster (5010 Steptoe St. at Boulder Highway, 702-451-6661) The oldest "private swinger's lifestyles house party" in Las Vegas. Notorious for just being the Red Rooster, it's both everything and nothing you'd expect. After 20 years of word-of-mouth only, the couple who've been hosting the parties finally broke down and put up a web site (www.vegasredrooster.com) that makes the experience more accessible.

Drive for a short while off the beaten path down Tropicana Boulevard to the Boulder Highway, and a few quick turns later you're in the manager's office of the Red Rooster Storage Facility. It's an actual storage facility. Inside, a man will ask you what you want, and if you don't talk about stowing boxes of junk for a while, he'll likely sell you a membership in his private club. Single men pay more and get one-night passes, couples get a week-long pass. Weeknights are slow. Weekends are insanely crowded.

Enter the 12,000-square-foot house of 20 or so bedrooms with a giant winding indoor pool, a full dance floor and stage for bands/DJ, a pool table, big-screen pornos, and every mix and match of swinger (or wannabe). Crowd ranges from mid-20s up to grandpa, and while there are plenty of unattractive folks lurking around, they do not outnumber the pleasant to look at.

Ultimately, the majority of people aren't too bizarre, but just out to have a good sexually charged time. Upstairs is a "couples only" area where people are getting it on. The action in this area might turn orgiastic (and sometimes it even happens in the general area), but at the basic level it's exhibitionists and swarming voyeurs taking their turns at doing what each does best.

The sheer volume of oral sex can be mind-blowing to the uninitiated, but that's what's goin' on here. Keep in mind the rigorous enforced "no creepy creepy" rules, which means no unsolicited touching, no knocking or going through closed doors, and no second requests for anything if the first was turned down. Lucky for most, this does not preclude leering.

In sum, a nice time for those who like to swing and enjoy random conversations with people interested in having sex with you. Or not. Bring your own booze, too: there's a "volunteer" who will mix it for you all night long. Red Rooster is also a good place to learn about the smaller lifestyle communities in the city as well as the offshoot parties that are hosted by members of the clan. Other good resources are www.swingers board.com and www.janesguide.com. 9 p.m. until 3 a.m. Variable cover charge based on event, and couple or single status.

The Green Door (953 E. Sahara Ave. Suite #B28, inside Commercial Center, 702-732-4656, www.green doorlasvegas.com) A private facility in a commercial center that sells memberships, patterned after the more homespun Red Rooster, but with less freedom to let your freak flag fly.

This place has three parts: a coffeeshop/internet-access space that's free to enter and affectionately referred to as "Starfucks" by its owners; the nuts-and-bolts main floor with pool table, sauna, lockable rooms, couples-only area and dark peek-a-boo alcoves for the male voyeurs who comprise the bulk of the clientele;

and the upstairs maze of lavish themed areas with more twists and turns than a John Grisham novel and equally complex as to what's supposed to happen there. Now and again, some couples decide to put on a show in the downstairs area as random guys hope for their chance in the big league to get their turn at bat.

Upstairs, with its crazy sex chairs and more comfy couches and non-secluded zones, is where most of the in-view hook-ups happen. Single guys have access, but it's massively more expensive. The whole place, if not super-friendly, tends to be super-busy, and the rumors of this being the stomping grounds for a bevy of attractive nymphos looking for all comers isn't exactly false … good luck on timing it right. Single men: $60 admission. Couples: $40. There are a number of time and access packages that start from there.

Club Oasis (and the rest) (700 E. Naples, at Swenson, no phone number) This place has gone through so many name changes and allegations of illegal prostitution it's a wonder it stays open, but as of print, Club Oasis (formerly the R&R Social Club) is fully in effect. There have been a lot of clubs like this one over the years. The concept is simple. You pay a cover price (typically $50) and get a tour of the facility by one of the many hostesses. The place is described to you as a "swingers club" suggesting that at any moment hordes of "swingers" will enter the facility and start "swinging" and may invite you to "swing" along.

Of course, that never materializes. In the interim (you know, just to kill time), the hostess would be perfectly willing to dance for you or give you a light massage. All you have to do is rent a room for a while.

Pay the fee. Once in the room, the game and additional costs for what's what are laid out.

After your chosen activity within the four walls is complete, you're free to sit on one of the couches and watch bad TV movies of the week until you either get re-inspired for another hostess session or the mythical swing party (lead by head swinger Godot) shows up. There are a number of other places in town that come and go proclaiming to be swingers clubs. All have large cover prices. Many are either Green Door imitators with smaller followings or Club Oasis places with less women working. All are even sketchier than their models. The ones that don't have hostesses may involve quite a bit of waiting around before any soul arrives. You may literally be the only person in the room for hours as most everyone who's actually into this scene is at the Red Rooster or the Green Door. 8 p.m.-6 a.m. $50 cover charge.

Warning: Clip Joints

On the one hand, while it might suck to pay a hefty cover charge to enter a place that might be completely empty, at least it's not a steady drain of cash. Not true for the clip joints or sex-tease clubs that have proliferated through the years throughout Las Vegas.

A quick history lesson: prostitution is legal in outlying counties surrounding Vegas. There's a city called Pahrump a mere 50 minutes away filled with brothels (see description, p. 174). Sometime in the early 1980s, a genius businessman decided to create an entire industry devoted to making people think they were in a brothel within Vegas. It worked brilliantly.

You would enter a place with a provocative name

like "Busty's" that had a sort of Old West whorehouse façade. The cover price was roughly $50. You would be swept to a counter where the overly-friendly bartender would make you his quick pal. The rap was scripted to the letter. Soon a line-up of working girls would fill the room, and you would be asked to choose whom you wanted. Cut to the chase, every component of the place was designed to make you depart with money with the implied guarantee that you were going to get laid if you just jumped through enough expensive hoops.

Under signs in bold print that clearly reprint applicable statutes explaining how really, really illegal prostitution is in Las Vegas, the bartender would suggest that buying a bottle of outrageously expensive non-alcoholic champagne was the way to get things started (as this wasn't a direct payment to the girl). In the room, the girls would have the patron "buy" them sexy clothes to take off and condoms at $20 a piece. Then, of course, once the money was gone or the guy tried to collect on the "promises," a burly bouncer would show up and toss the man to the curb. What could he do? Complain to the police that he didn't get the illegal prostitution he was entitled to?

There are legends of horny computer tycoons spending upwards of $10,000 for a handful of condoms and some dirty talk. Eventually, the sex-tease magnate tired of the feeble efforts of vice to shut him down, and the clubs went away. That is, until others thought it would be a shame to let such a good idea go to waste. Almost indistinguishable from the "swinger club/secret brothel" concept, these places are most apparent by the requests for multi-levels of expenditure to get to the non-existent "goods." To remain legal, they offer some service.

Past clubs have offered spas, "tickle" service and the like, but once you're in the private area, if no one has told you in non-ambiguous terms that you will get sex (not "feel good" or "get what you want"), then this is *not* a sexual business. It's a sex-tease business, and unless you like what you're getting at that point, understand that it's going to stay the same, only more expensive.

Las Vegas Call Girls

Walking down the Las Vegas Strip, it's hard to avoid the packs of ESL (English as a Second Language) handbillers thrusting glossy cards with scantily clad women who will come to your hotel room and "fulfill your fantasies" for some outrageously low sum. Sometimes it's twins, sometimes women with breasts the size of Thanksgiving turkeys. Looking around, you may see billboards or cab tops touting the same service with a silken-lipped beauty promising to be there in 20 minutes or less. (Hey, she can beat the pizza!)

Then, you walk into the casino, and while most of the women seem scandalously dressed in the faux-hooker fashions of the moment, some appear to be actual hookers. Are they? Finally, up in your room you pop the five bucks to get some internet access and you just accidentally Google the words"hot Las Vegas barely legal sex." A number of sites pop up. What do you do?

Handbill Escort Agencies

The little card or pamphlet handed to you on the street likely comes from one of the larger escort agencies in town. The woman on the card existed only in some photo studio in West Hollywood circa 1987. The number leads you to a pleasant-voiced women whose job is to assure you that this is the normal way for guys to get laid in Vegas without saying it outright.

The rap goes something like this: "What kind of girl would you like? Oh, I can send a woman just like that over to your room right now. The fee (from $69 to $300) is just for her to come over and get completely naked and wild with you, that's the agency fee, anything else is strictly between you and the girl."

This dialogue is necessary to shield the agency from allegations of prostitution. Once she's certain you are not law enforcement, she'll ask about your budget, or give you some sticker shock to start: "My basic show starts at $1,000." A vast majority of these women are prostitutes looking to get maximum value for minimum time. It's all a crapshoot, and you may or may not get anyone remotely resembling what you requested. Field operatives report that the going rate for basic sex acts is $150 on top of the agency fee. A number of agencies also advertise in the back of alternative-weekly magazines.

Call Girls Trolling the Casino Bars

Because the casino industry is highly regulated, casino security is required to show some effort to prevent call girls from setting up shop on their properties. These efforts are either woefully inadequate or lacking by design since on any given night there are hundreds of women working the floors. Actually, most are there only on layovers. With cell phones prominently in sight, many of these women are agency girls waiting for a call and looking to possibly pick up some independent business for the downtime. The key to picking them out of the crowd is their eye contact with every man, saying "hello" or "where you going?" as you pass by.

If you stop, there may be what seems like general flirting and small talk. You may even wind up buying her a drink. Actually, you may be a stud who has scored a hot babe in a swingin' casino. More likely, you are being sized up, and soon they'll reveal their status and price. Typical range is $300 to $700. Their prime directive, however, is to get out of there to take the next phone call that comes through, so the encounters are typically rushed.

This is in comparison to the street prostitutes who are commonly arrested by the police and who'll offer things you didn't know could be done in America for $20. Good luck. Oh, and by the way, the police with time on their hands tend to run reverse-stings every now and again on the streets to get the johns. In an interesting view on the nature of the business, the street hookers wind up getting serious time, while the johns are given the chance for "reform" by attending classes where they are forced to watch (like Alex in *A Clockwork Orange*) giant blow-ups of diseased ravaged genitalia. Yikes!

Internet

Oh, Hallelujah for technology. What better way to match "provider" with "hobbyist"? The Internet has become a good way to locate (and then screen for price, appearance and services) what you are looking for.

There are a number of sites run by agencies, but others that seem filled with independent "provider" listings. Popular websites include the Eros guide (**www.eros-lasvegas.com**) and Vegas Exotics (**www.vegas exotics.com**). Listings include women for men, men for women, men for men, and all the variations in between.

Many independents have their own websites and quite a few give a link to "The Erotic Review" (**www.theeroticreview.com**). This service exists in many cities, but because of the vast number of sex workers based in Las Vegas, there are literally hundreds of escorts, masseuses and bondage professionals with detailed information about what happened and how. While general facts are available for anyone who logs in, the particulars are reserved for those who pay the modest monthly fee. Still, there's a disclaimer that all the encounters are fictional, yet the ranks of the eager to take a chance on well-reviewed "providers" seems to grow. A lot of it is given in code: "Porscha caught me with a surprise DFK, which led to us removing our clothes. We fell on the bed and to my delight she started a BBBJ, which was mind blowing (if not other parts). I returned the favor when I decided to DATY. Then she slipped on a cover and hopped on for some spirited CG and RCG. I rolled her off for the MISH and while I tried to speak to her in GREEK, it was clear that it was a foreign language. Oh, well, I did what I had to do, spilled a cup, cleaned and cuddled. Great time." The Erotic Review also has search features that allow users to find the most popular providers. Ah, technology.

Brothels

Sheri's Ranch (Next to the Chicken Ranch, where Homestead Rd. deadends; 1-866-820-9100, www.sheris ranch.net) Sex for money is absolutely legal 50 minutes outside of Las Vegas. It's touted as cleaner, safer and more satisfying than the illegal prostitution that's readily available in Vegas. (Indeed, industry lobbyists—yes, they have lobbyists—repeatedly cite the regular sexual-disease exams on workers and not one case of AIDS or HIV to date.)

Now and again, local citizenry fueled by righteous indignation and a message (not massage) from the Heavenly Above launches a protest to rid the towns of demon brothels, but usually to no avail. On the outskirts of Pahrump, Nevada, is the closest and nicest brothel. Sheri's Ranch is the Taj Mahal of legal whorehouses, though that's not saying much since most have been stuck in "connected trailer mode" since the legalization of the trade in the 1970s.

Sheri's has reinvented the form by providing a user-friendly brothel for the new millennium. Gone are the intimidating gates with buzzers and austere madams laying down the business vibe. Enter an attached sports bar frequented by locals (men and women) who have no interest in the services other than a stiff drink at a reasonable price.

For those more intrigued by the sexual offerings, there are two ways to go. First, just strike up a conversation with a working lady at the sports bar, and she'll take you to her room for discussions of price and service (discussing these matters, even between patrons, is *strictly taboo*). Second, you can request a traditional line-up to make your pick done in front of a large window, which allows the natural light (sun or moon) to shine through.

The minimum starting price for services is about $200, though not unlike their illegal cousins in the city, they'll size you up for what you might be willing to spend. Some packages go up to $10,000, generally offering the life of a sultan for 24 hours.

Sheri's also offers full spa treatment, themed rooms, a full-sized pool and eventually an 18-hole golf course. Oh, and sex in all its wild variations. Rule of thumb is the more you spend, the more things get done, but like anything else, it's a business, and they'll try to give you the least for the most. All in all, Sheri's has successfully created a more upscale version of what has been a low-rent industry. Think of it as a sexually-charged Doubletree straight out of Mesa, Arizona. Oh, and as a slight twist on the theme, there are options for couples and even single women patrons to engage in some good old-fashioned legal sex trade. 24 hours. No cover. Bar.

Mabel's, etc. (US 95 North to Highway 160, look for the signs and turn right at Crystal, Nevada. 775-372-5469) Before Sheri's hit the scene in Southern Nevada, Mabel's was queen bee with the others, like the **Chicken Ranch** (10511 Homestead Rd., Pahrump, 1-877-585-2397. 24 hours. No cover. No bar. www.chickenranchbrothel.net), lagging not far behind.

Still, these old warhorses in the battle for brothel supremacy manage to trudge along offering legal prostitutes that aren't much less appealing than Sheri's and at a bit more affordable prices. Because all the brothels in the area are not far from each other, a discriminating shopper could make the rounds, get a line-up, and come back to the one stand-out sure-thing that made the trip through the desert worthwhile. That is, of course, if the shopper was in charge of transportation. Cab rides get expensive and limo service, while free, was probably provided by the brothel that owns that limo. Otherwise, pick a place and have fun. 24 hours. No cover. Bar.

Massage Parlors

Asian massage parlors have popped up like toadstools along a few major thoroughfares, and all reportedly offer the same routine. Around $40 for a half hour massage, $70 for an hour. You are led to a room by the person who took your money and told to strip naked and lie on the table. That person is going to be your masseuse and likely has just arrived in America from lands of the Far East. English will be a barrier, but various international signals can always be relied upon.

A clinched fist slowly rising up and down indicates that a handjob is available. The thumb gliding across the index and middle finger making a scratching sound indicates that money needs to be offered before the famed "happy ending" is to occur (general price range is $40 to $100 more). Keep in mind that before the high octane fun begins, many places offer real massages that are easily worth the reasonable fees.

While one place may be virtually indistinguishable from the next (in fact, they sometimes tend to share workers), one that takes a more upscale approach and has a reputation for good work on the deep-tissue rubbing side of the coin is **Ocean Spa Massage** (6020 W. Flamingo Rd. at Jones, 702-257-8800, 11 a.m.-5 a.m.), with locker areas for changing, full showers available, and rooms equipped with those metal racks on the ceiling which sets up the whole barefoot walking treatment.

Adult Bookstores / Peep Shows

Adult Superstore (Four locations: "Mega-store" located at 3850 W. Tropicana Ave. and Valley View, 702-798-0144. Open 24 hours.) There probably are a few independently owned adult-oriented bookstores left in Las Vegas, but you'd be hard pressed to find them since the Adult Superstore has either bought up or crushed the competition.

The Wal-Mart of Pornography, the Adult Superstore has somehow managed to be all things to all smut shoppers. Discerning couples perusing just the right double-headed marital aid might be browsing right next to the overcoat fellow feeling just how realistic the money part of the Jenna Jameson blow-up doll is while an entire bachelorette party is sizing up penis hats. Meanwhile, in a more secluded part of the store is an arcade showing hardcore flicks for the lurking quarter-jiggler. The best and biggest Adult Superstore (and least seedy) is on Tropicana road just east of the Orleans Hotel.

Showgirl Video (631 Las Vegas Blvd. S. at Garces, 702-385-4554) The only porn retail, hardcore video arcade, and accessory shop with an almost around- the-clock peep show in the back. Sometime in the wee-morning hours workers close the ammonia-laden booths, hose down the glass windows, and empty out the automatic bill validators. It is what it is. Fully naked women getting up close and personal against a sheet of fogged Plexiglas that clears upon entry of a dollar into the slot, then disappears a minute later (until the next dollar goes in). Gals work for tips. Still, that doesn't seem to stop them from being completely detached from the wiggle-and-reveal show they are the stars of. It's not uncommon to see a fully nude woman on the cell phone or filing her nails while the patron is furtively feeding dollars with the one free hand. All this makes sense if you understand the presumptive motto of Showgirl Video—*what happens in your booth, stays in your booth … until the mop man knocks twice.* (There's a rumor that this inspired its more famous ad campaign cousin for the city of Las Vegas.)

Oh, and if you like the girl in the window, you can go to a one-on-one booth replete with a telephone to talk to her through a much larger unfoggable Plexiglas. Show-and-tell in the back booths start at $25. Bookstore open 24 hours. Dancers off from 5 a.m. until 8 a.m. $1 cover.

WHAT CULTURE?

Vegas Art Attack

It's when you stumble out of one of those mall-sized "gentleman's cabarets," slightly drunk and disheveled, fingering your pockets for that last $20 you stashed and wondering where the hell all your friends disappeared to: that's when they'll come for you. The vintage limousine will glide into your path as silent as a shark. Before you're aware of how exactly it happened, you'll find yourself in the back seat, surrounded. You'll dimly wonder why all these guys are wearing white linen suits out of the 1940s, staring at you intensely through their glasses, their Panama hats cocked at rakish angles. In the initial silence, you'll feel fear creep up your shooter-addled spine. At last you'll speak:

"Um, what are you guys, the metrosexual mafia or something?"

Icy stares, and one chuckle from the darkened seat opposite from you.

"You're a funny guy, Clyde," says a voice. "Just the other day, I was saying to Vinnie—and I had to say it loud, 'cause, you know, Vinnie's only got the one ear—I was saying, 'Vinnie, what we need is a funny guy for our little ride.' Looks like you fit the bill, pal."

You'll gulp, a cold sweat beading on your forehead. "A little ride? Uh, where?"

Chuckles all around now. "Why, to the Las Vegas funny guys like you never see, pal," says the Voice. "The Vegas that you might see if you could get out from under a lap dance long enough to see it. The Vegas that struggles to make beauty while you poke your buddy in the ribs because Ben Affleck is losing his shirt at the next blackjack table. The Vegas unseen but by those whose eyes are open, that creates its own world inside the long shadow of the casinos and the sub-divisions." And then the Voice leans forward and fixes you with a menacing pair of eyes lit with a dark fire. "The Vegas of art and culture, funny guy. Pablo, read this guy some chapter and verse."

In a thick European accent, someone else in the limo begins to recite: " 'One cannot imagine public "art," let alone a museum, on the Vegas strip. It would have nothing to do there except look high-minded and insignificant. Here the idea of art simply evaporates, it flies off in the face of the stronger illusions with which this place is saturated: sudden wealth, endless orgasm, Dean Martin … Vegas sums up a certain kind of American giganticism ... not because the place is big, but because (at least at night) it seems to be. Its monuments, the city lights, are conceived on a scale far beyond anything that most artists ever get to work on. The town is a work of art: bad art, but art all the same.' "

The Voice snarls at you: "That's by some Aussie art critic named Robert Hughes. You buy any of that, Clyde?"

"Uh ... well, uh, aren't there museums on the Strip now?" you stammer, conscious now of the big shoulders you're squeezed between, the whiff of Paco Rabone strong in your nostrils. A Countess Mara tie dangles in the passing streetlights as the Voice leans forward.

"Smart and funny. You just might survive this little ride, pal. We got museums, and we got an art scene, and we got more icons than that mouthy fuck Hughes can shake a wattle at. Las Vegas is a work of good art, pal, and tonight you're going to see the people and places that keep it that way."

That's when you almost make your last mistake. You snicker a little, and the words come out before your brain can stop them: "Oh yeah, sure. Everyone knows there's plenty of culture in Vegas."

A meaty hand closes around your throat. Spots dance around your eyes as another voice thick with menace penetrates your oxygen-starved brain. "You want your body parts to be appropriated for a mixed-media collage, funny guy?"

"That's enough, you big dumb Pollock," says the Voice. "Funny guy is beginning to see the light. Aren't you?"

You nod and gasp, and that's when your real night in Vegas begins. The limo heads for downtown, roaring past a building with a big, classic 1950s neon clock and sign that reads HOLSUM BREAD ... *hours fresher.*

"See that old bakery, Clyde?" says the Voice. "Every night you'd drive by and smell the fresh baked bread. Then those fucks moved their bakery out to Henderson. But you know what? That building ain't coming down. They're turning it into the **Holsum Lofts**, full of studios and galleries and live/work spaces for artists. 'Cause this, my friend, is the Arts District."

The limo pulls up to Charleston and Main, and that's when you find out about it all: how on **the first Friday of every month** (www.firstfriday-lasvegas.org) the renewal of downtown Las Vegas continues, with the whole Arts District open all evening long. How the folks from the city's only historic district, **the John S. Park neighborhood**, turn out along with all the art-starved zombies stuck out in the 'burbs to check out the galleries and music and performances. How they go to the two-story **Arts Factory** (101-109 E. Charleston Blvd. at Main, 702-676-1111, www.theartsfactory.com) and check out the city's only artist-run non-profit, the **Contemporary Arts Collective** (702-382-3886, www.cac-lasvegas.org), or move and shake things upstairs amidst the abstract fields of the **Michael Wardle Gallery** (702-383-8633, www.michaelwardle.com), or see what's cooking with the group of urban artists in the **Five Finger Miscount** studio. How they go next door to the **S2 Art Group print shop** (1 E. Charleston Blvd. at Main, 702-868-7880, www.entertainmentgalleries.com) and gawk at the 19th century presses that turn out fabulous lithographs of famous posters.

"But here's the best reason to go to the Arts

Factory," says the Voice as he pushes you into **SEAT**, the theater that's home to **Test Market** (702-736-4313, www.godsexandbowling.com), the city's best avant-garde theater company and space. "While you were trying to get you dip your wick in some over-priced call girl at the piano bar in the Bellagio," the Voice says through a cloud of cigar smoke, "we were getting our intellects stimulated by productions of *Happy Days*, *Caligula*, *Fool for Love* and *Equus*." The Voice pauses. "Not that getting your wick dipped and innovative theatre are mutually exclusive."

"*Caligula* was actually put on by that groovy guerilla theatre group, boss, called the **Cockroach Theatre** (www.cockroachtheatre.com)," says Vinnie.

"You're gonna feel like a cockroach if you interrupt my flow," says the Voice.

The suits push you out onto the street, down Main where they show you the city's most cutting-edge contemporary art gallery, **DUST** (1221 S. Main St. just south of Charleston, 702-880-3878, www.dust gallery.com). "This is the place, Clyde: the cream of Vegas's artists, all those kids who studied with Dave Hickey at UNLV in the 1990s, show here, along with artists from New York and Europe." The Voice fixes you with a stare. "You do know who Dave Hickey is, don't you?"

"Of course!" you'll squeak out, desperately trying to keep your knees from buckling as your escorts grab your arms and drag you down to Casino Center and Colorado, where the fabulous **Funk House antique store and gallery** (1228 S. Casino Center Blvd. and Colorado, 702-678-6278, www.thefunkhouselas vegas.com)—nerve center of First Friday—is just across the street from the little galleries and studios, like **Dray's Place** (1300 S. Casino Center Dr., Suite 5, at Colorado, 702-255-9674) are taking over the run-down apartments and transforming the streets into a cultural center.

"It ain't all paintings, right, boss?" says one of the mob, who turns to you with a sneer. "We got poetry downtown, too, laughing boy."

"That's right, Andre," says the Voice. "Let's take our pal here to the Bridge."

"No, no, I'll do anything you say!" you cry out on the edge of hysteria.

"Relax, Clyde. A fall from the **Poet's Bridge** won't kill you—unless Mayor Oscar Goodman's river is full of gin." The suits laugh all around, but you don't get the joke until they take you down to the **Lewis Avenue Corridor park** (between Las Vegas Boulevard N. and Fourth) and show you: a concrete bridge inscribed with lines by 20 Nevada poets above a faux creek bed, now bone-dry.

"The Mayor had water running through here," says Pablo mournfully. "But there's a drought emergency now, you know."

"Maybe water, or maybe gin!" The suits laugh as you nod and smile, relieved.

"You thought literary Las Vegas was all fear and loathing, didn't you, Clyde?" You can only smile weakly, your head spinning. "Nah, we got us plenty of talented writers here. Poets come in from out of town all the time and you know what they say? That we got the best audiences! Can you beat that? You can check out who's reading on the **Literary Event Calendar** (www.localendar.com/public/VegasPoetry). Plenty of open mics and features. We even got a nice little bookstore in a casino now, pal. Maybe you noticed the **Reading Room** there in Mandalay Bay (3950 Las Vegas Blvd. S. at Mandalay Bay Dr., 702-632-9374) when you had to break away from the slots for a piss."

"I dunno," says Vinnie. "Maybe Mr. Funny Guy here only reads funny books. You didn't know one of the best comic-book stores in the West is right here in Vegas, did you pal? **Alternate Reality Comics** (4800 S. Maryland Pkwy. just north of Tropicana and across from UNLV, 702-736-3673, www.altrealitycomics. com). Maybe Frank Miller's *Sin City* comics is more your speed." Vinnie leans in and leers at you as if Frank Miller himself had just scratched out his features with a bloody pen nib.

"Our pal here is looking a little pale," says the Voice. "Let's brighten him up with a nice, long walk. We can walk right up Las Vegas Boulevard, underneath all the artists in the **Aerial Gallery**, all those fabulous vinyl banners featuring the works of a dozen Vegas artists hanging up on the old streetlamps from Charleston to Fremont."

A big arm pushes you forward, toward the **Fremont Street Experience Canopy**. "That's not the real stuff," says the Voice, looking up at the four-block-long video screen above Glitter Gulch. "Over here, pal: examples of Las Vegas's most original art." They lead you up and down 3rd Street and Fremont (at Las Vegas Blvd. S.) to see the **Neon Museum**, restored signs from the city's glory days: the original Aladdin's lamp, the caballero from the Hacienda, Vegas Vic greeting you atop the Westerner Motel sign. "Of course, these are just teasers," says the Voice. "The real Neon Museum is gonna go up down the street, where the Neon

Boneyard is, full of signs just waiting to be restored."

"Let's take laughing boy to the Boneyard, boss," giggles one of the suits.

"Not tonight," says the Voice. "**The Neon Boneyard**'s not open to the general public yet. You have to make an appointment. Have you got an appointment, smart guy?"

"N-No..." you say. "But I'll make one! Anything you say!" You stumble along the sidewalk amidst them, fully sober now, terrified.

They laugh, harsh as the neon in your eye after an all-night binge. "It ain't all happening downtown, Clyde," says the Voice. That's when they shove you back into the limo and take you out north to the **Left of Center Gallery** (2207 W. Gowan Road at Martin Luther King, 702-647-7378, www.leftofcenterart.org), with its fabulous community art space dedicated to African-American art. That's when they go down by the university to the **Gallery Au Go-Go** (4972 S. Maryland Parkway, Ste. 11, just south of Tropicana, 702-419-5681), where the tattoo artist/punk musician Dirk Vermin runs the most rock 'n'roll art space this side of Death Valley.

That's when they drive you by the truly hidden treasures: **Castillo del Sol** (www.principality-of-paradise.com/index.html), the crazy, Addams Family-style home of brain surgeon and former Lt. Governor Lonnie Hammargren ("Sometimes he opens it up for an ice-cream social," the Voice tells you) and the **Pinball Hall of Fame** (702-434-9746, www.pinballmuseum.org), where twice a year you can play pinball on 400 vintage pinball machines all night long ("For charity, mind you," the Voice admonishes).

The limo glides along the streets. "You getting the point now, Clyde?" says the Voice. You nod furiously. "You can even keep up with what's happening with the art scene online at Robert Kimberly's **LV Arts and Culture Blog** (http://lvartsandculture.blogspot.com). Got all the links a smart guy like you needs to stay in touch. You can work a computer, can't you, smart guy?"

"Sure!" you cry. "I can stay in touch! Thank you, I never knew any of this was here, thank you, thank you, thank you so much!" You're babbling now, the hysteria creeping back up. "And hey, I won't tell a soul about tonight! What happens in Vegas stays in Vegas right?" You laugh, on the edge of panic.

Things get quiet. They shake their heads. "You ain't such a smart guy after all," the Voice says softly. "We want you to tell the world, pal. Tell 'em about the other city of lights." He sighs. "Looks like we've failed, boys. We're gonna have to take an even longer ride." Silence fills the limo like a cloud of poison gas. "A long ride out to the desert."

"No ... NO!" you cry out.

"Let's go out to Beatty," says Vinnie. "We can show him the sculpture gardens and the ruins of the Rhyolite mining town at the **Goldwell Open Air Museum** (2.5 miles west of Beatty, Nev., off of State Highway 374, 115 miles north of Vegas, www.goldwellmuseum.org). It's only a couple of hours away."

"Plenty of spots to dig a hole out there," mutters Pablo.

The Voice is silent. "We don't have time for that. Blindfold Mr. Funny Guy."

They blindfold you as you plead for your life. After a few minutes, the limo rolls to a stop. They drag you out, and you stumble up some steps, your heart beating as if it's about to jump out of your chest. You keep hoping your life will flash before your eyes, but all that's coming up are episodes of "American Casino." You sob as the Voice calls out, "Right here. This'll do."

Hands force you onto your knees. You're begging now. "I get it," you sob. "Vegas has all sorts of art and hidden culture! I get it! Please don't kill me!"

A few chuckles, then silence. The Voice sighs again. "What can I tell you, pal? *Ars long, vita brevis.* But hey, you're up more than if we let you wind up back at the craps table, right?"

Something cold and long is stuck in your mouth. You close your eyes tight, and wait for it. After a while, you realize nothing has happened. You pull the blindfold off and find you have a paintbrush stuck in your mouth. You're sitting at the foot of huge, dark steel object in the middle of some plaza. There's a ring of light around the base of it.

You stand up slowly, looking up at it, then out across the campus of the **University of Las Vegas-Nevada** (4505 S. Maryland Pkwy. at Tropicana, www.unlv.edu) as dawn lays her rosy-fingers across your shoulders.

"**Claes Oldenburg**," you whisper, looking up at the sculpture. "*Flashlight*, **1981**. The city's best-known public sculpture."

How about that? You're practically a made man.

Need Money ... Yesterday?

Once the casinos have had their wallet-thinning way with you, **SuperPawn** (2300 Charleston Blvd. at Fremont, 702-477-3040; 1611 N. Las Vegas Blvd. at Main, 702-642-1133; 2645 S. Decatur Blvd. at Sahara, 702-871-4464; 3270 S. Valley View Blvd. at Desert Inn, 702-643-9851; 4635 W. Flamingo Rd. between Decatur and Arville, 702-252-7296) is the most consistent place to sell things like your wedding ring, your grandmother's fancy heirlooms, your ex-beau's PlayStation, or whatever you've got lying around that's at all attractive. What's meant by "consistent" is that

SuperPawn won't lowball you too badly, but neither will they be generous. (Independent pawn shops, on the other hand, are sticklers and will give you nothing but the absolute minimum.) Also, if you're in the mood for buying something on the cheap, you're more likely to find a good deal at a SuperPawn than at any other pawn shop, according to those in the know.

Unloading musical instruments is another matter, though. While SuperPawn won't completely rip you off, the best place to sell your sonic gear is at **Cowtown Guitars** (2797 S. Maryland Pkwy. at Karen, 702-866-2600). This place specializes in vintage instruments, so the stuff on display is a bit pricey. But if you just need a cheap Japanese six-string to beat up on, just go SuperPawn.

And if you're really down on your luck, you can sell your plasma. There are many rules involved: most "plasma donation centers" require proof of residency for seven days (hold on to those hotel receipts!), a picture ID, your Social Security card, and you can't be an intraveneous drug abuser or a practicing homosexual. Also, you can't have enjoyed alcohol within 24 hours of giving plasma.

You'll get the most buck for your blood at **BioMat USA** (611 N. Las Vegas Blvd. at Bonanza, 702-385-5172), where they'll give you $30 at each of your first two visits, and then after that, $25 a pop for up to two visits each week. Get up early to sell your plasma, though: BioMat USA is open every day except Sunday, 8 a.m. until 5 p.m. But they only accept new donors before 2 p.m.

IBR Plasma Center (1912 Civic Center Dr. at Lake Mead Blvd., 702-642-4037) gives you a little less money for your plasma ($25 for each of the first two visits, $20-25 thereafter), but not much less. Also, be sure and call ahead, since their hours are spotty.

And then there's **Pyramid Biological Corporation** (1732 Fremont St. at Bruce, 702-385-7337). Here, you need proof of 60 days of residency in Vegas. Pays on par with IBR Plasma Center. Open weekdays, 6 a.m. to 8 p.m.; weekends, 7 a.m.-3 p.m.

Wanna jack off for money? Sorry, not in Las Vegas. The nearest sperm bank is in L.A.

About the Contributors

Editor **Jarret Keene** is a poet, professor, musician, journalist, and author of the poetry collection *Monster Fashion*. Keene's prize-winning poems, stories and essays have appeared in more than 100 literary journals. He serves as A&E editor for *Las Vegas CityLife* and teaches creative writing and literature at the University of Nevada, Las Vegas. His indie-rock song "Superbleeder," from the album *Die Kinder* (Mperia.com) appears in the Z-grade splatterflick *Blade of Death*.

Seth "Fingers" Flynn Barkan's *Blue Wizard Is About to Die!: Prose, Poems, Emoto-versatronic Expressionist Pieces About Video Games, 1980-2003* is the first collection of poetry about video games ever published. Barkan is *Las Vegas CityLife*'s bar critic, and his ultra-violent video-game column Kill Everything is soon to be nationally syndicated. He's also the resident pianist at the Freakin' Frog Beer & Wine Café.

Gregory Crosby's verse has appeared in numerous literary magazines such as *South Carolina Review*, and his poem "The Long Shot" is forever immortalized in bronze on the Poets' Bridge in downtown Las Vegas. He has written for numerous travel publications.

Scott Dickensheets' illustrations have appeared on napkins in many of the Southwest's finest restaurants, as well in such publications as *Las Vegas Life* magazine, the *Las Vegas Weekly* and the *Las Vegas Sun*. He has received a Best Illustration award from the Nevada Press Association. Now editor of the *Las Vegas Weekly*, Scott, a 30-year resident of Southern Nevada, lives in

Henderson, with his wife, three sons, two dogs, two cats and a houseful of doodles.

Joshua Ellis is a writer, rock star and web guru. His weekly column All Tomorrow's Parties appears in *Las Vegas CityLife*. You can save your soul at his website, www.zenarchery.com.

Real-estate agent **Izzy Fizler** has documented his Sin City sexcapades with a host of lessor-known celebrities of the 1970s in his upcoming memoir, *Saturnalia—My Life on Mars*.He enjoys the music of Bob Wills and is an occasional submitter to *Screw* magazine under the pseudonym, "Tiger Boy."

Anne Davis Mulford, aka Princess Anne, was once a gossip columnist for the gay press in Las Vegas. She's now a renowned sculptor and visual artist.

Jennifer Prosser serves as the editor of the Las Vegas-based *Monorail* magazine. She lives with three cats and a turtle.

INDEX